Earth Science

PHIL MEDINA

Homework Helpers: Earth Science
Edited by Kathryn Henches
Typeset by Eileen Dow Munson
Cover design by Lu Rossman/Digi Dog Design
Printed in the U.S.A. by Book-mart Press

To order this title, please call toll-free 1-800-CAREER-1 (NJ and Canada: 201-848-0310) to order using VISA or MasterCard, or for further information on books from Career Press.

The Career Press, Inc., 3 Tice Road, PO Box 687,
Franklin Lakes, NJ 07417
www.careerpress.com

Library of Congress Cataloging-in-Publication Data

Medina, Phil, 1967-
Homework helpers. Earth science / by Phil Medina.
p. cm.
Includes index.
ISBN 1-56414-767-3 (pbk.)
1. Earth sciences. I. Title.

QE28.M43 2005
550--DC22

2004058484

Acknowledgments

Special thanks to my wife, Valerie, for her support and help she has given me while writing this book; my parents for opening every door that I ever wanted to walk through; Jessica Faust for offering me the opportunity to write this book; and John Hotz, the Queens Village Clockmaker, for opening my eyes to the wonders of nature and science that are all around us. In his basement clock shop I was the apprentice to DaVinci.

Thanks also to my high school French teacher, Gail Stein, for showing me that a true teacher strives to be a master of her subject; and to Becky Arthurs, Jon Quigley, Annette Goldmacher, and all of my students who inspire me to learn as much as I can and to be the best teacher I can possibly be.

Lastly, thanks to my colleagues at Murphy Junior High for surrounding me with experts from every field and allowing me to thrive. I would especially like to thank Pam, Conrad, and Barbara for their help with this book.

Contents

Introduction

Welcome to Homework Helpers: Earth Science

Homework Helpers: Earth Science was written for the students who need a different presentation of Earth Science than what they've seen in class. This book is written less as a textbook and more as a conversation between the reader and a tutor.

Earth Science is the study of our natural surroundings of which we all have some familiarity. However, the important connections are often lost in the dry facts and abstract concepts. The discussions and examples in this book will give the reader the "oh, of course! That makes sense," reaction.

This book is not to be read in one sitting or even in the order that it is presented. It is written as a supplement to your classroom experience. Pick up this book and review each topic as you work on it in class or in preparation for the chapter test. If you are finding your class particularly difficult, it will also come in handy to preview the topic before you see it in class.

Each section has practice questions and exercises designed to reinforce the new material. After the practice sections, you will find the answers and explanations of correct as well as incorrect answers. At the end of each chapter, there is a chapter test also accompanied by explanations of the answers.

Earth Science is something we are all immersed in every day and should not be a foreign subject. A firm grasp of the concepts of how our world works makes us partners in a relationship with nature, rather than victims of it.

Introduction to Earth Science

Lesson 1–1: Observation and Inference

An **observation** is any information that is gathered by using any of the five senses. Observations are the facts that are used in science. They tend to be dry facts, measurements, and statistics. Think of observations as evidence.

A **measurement** is a form of an observation. This is referred to as "comparing to a standard" because the definition of the unit you are using, such as a meter, is a standardized distance. In order that all measurements agree with each other, the definition of that unit can be found somewhere. For example, in the state Bureau of Weights and Measures, there is a metal bar that is measured very accurately to be as exact as humanly possible to a meter. This bar is the "standard meter." All other meter sticks are compared to this one bar for accuracy. Whenever you use a ruler to measure a distance, you are "comparing to a standard." Even though your 49-cent ruler may not be the most accurate, it is a fairly close representation of that bar in the Bureau of Weights and Measures. By the way, this is a huge leap from the old days of measuring a cubit as the length from your elbow to the tip of your middle finger or a foot as being the size of the king's foot.

Some Examples of Observations

The Sun rises due east on March 21.

The rock has a hardness of 7 on Mohs scale of hardness.

There is a quarter Moon in the sky.

An **inference** is a conclusion that is made based on your observations. It will typically answer the question "why" or "how." When you make an inference you put the evidence together and explain what happened or what will happen.

Some Examples of Inferences

The rock was transported by a glacier.

This sample of granite cooled slowly from magma inside Earth.

Based on today's weather trends, it will rain tomorrow.

Lesson 1–1 Review

Identify each of the following statements as either an observation or an inference:

1. The rock is dark brown.
2. The Moon is 386,000 km away from the Earth.
3. The Moon was created by a collision between Earth and another object in space.
4. It's raining today.
5. According to the graph, the world's population will double in 100 years.
6. A glacier gouged out this valley.
7. The wooly mammoth was hunted to extinction by primitive humans.

Lesson 1–2: Patterns of Change

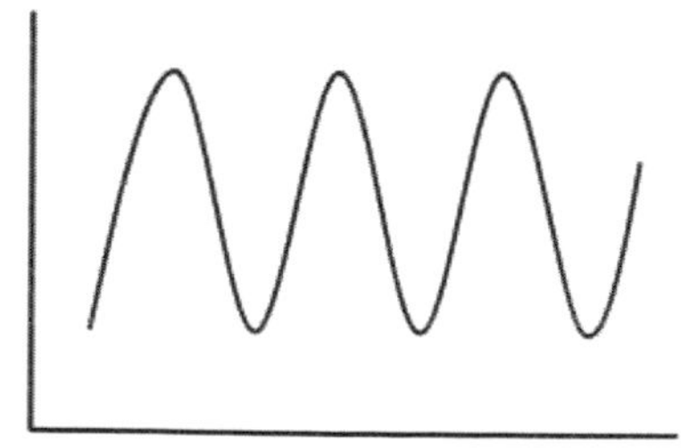

Figure 1.1. Cyclic Relationship

In Earth Science we study many kinds of changes: changes in temperature, changes in the size of rocks as they wear down, the changing position of Earth throughout the year, and so on. Many of these changes follow some kind of pattern, which allows us to predict what will happen in the future.

A **cyclic pattern** is a repeating pattern. This is caused by repeating events and, when graphed, it appears as some version of a sine graph.

Examples of Cyclic Patterns

The hot and cold of the day and night.

The hot and cold of summer and winter.

The phases of the Moon.

The rise and fall of the tides.

A **direct relationship** is a pattern of change in which the cause and effect both increase together or both decrease together. The graph of a direct relationship will slope upwards towards the right.

"As X gets bigger,
Y also gets bigger."

An **inverse relationship** is a pattern in which one variable increases in value while the other variable decreases. When this relationship is graphed, it will make a line that slope downward.

"As X gets bigger,
Y gets smaller."

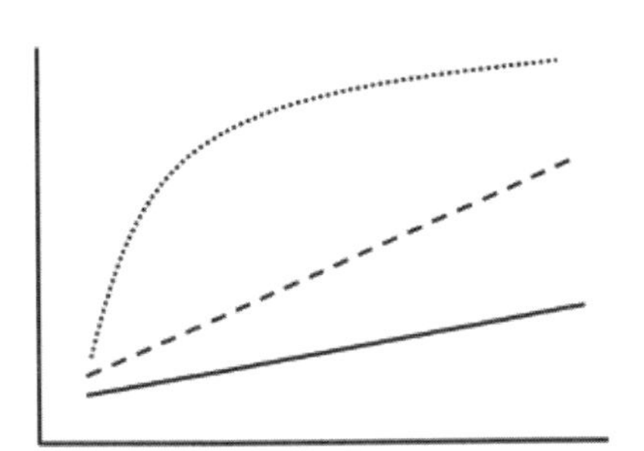

Figure 1.2. Direct Relationships

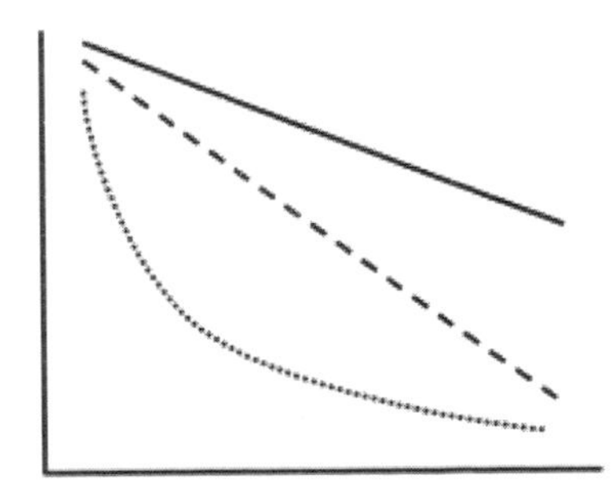

Figure 1.3. Inverse Relationships

To **extrapolate** is to take a pattern and extend it beyond the current data. Extrapolation is used to make predictions. A classic example of this is to predict the world's population in 20 years. On a graph, extrapolation is as easy as extending the line or pattern.

Interpolation is another way of using data to make a good guess about a missing portion of information. Contrary to *extra*polation (*extra* meaning "outside" or "beyond"), *inter*polation (*inter* meaning "inside" or "between") does not extend the graph. It "reads between the lines." For example, if you measured your height in fourth grade and you were 3 feet tall and in tenth grade you were 5 feet tall, even though you didn't measure yourself in seventh grade, it would be a good assumption that exactly between the fourth and tenth grade, you were exactly between 3 and 5 feet tall—or 4 feet tall.

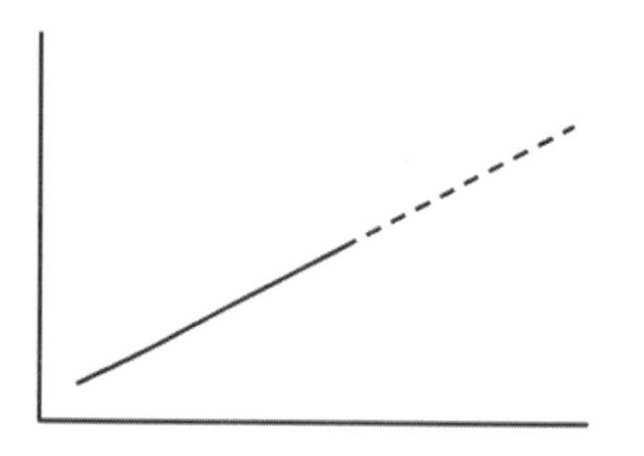

Figure 1.4. Extrapolation. To extrapolate a graph, simply extend the line.

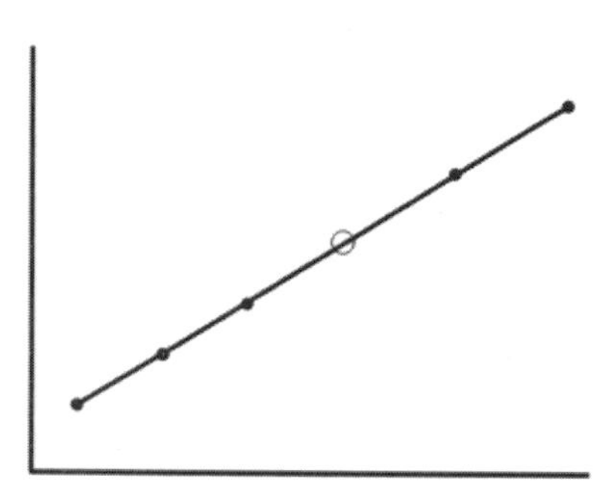

Figure 1.5. Interpolation

Making Graphs Without Using Numbers

Sometimes to show relationships graphs without numbers are used. But to make a numberless graph from scratch can be difficult until you know how to do it.

As an example, we'll use the relationship between temperature and density of a material. The first thing that you need to do is to label the axes: *temperature* and *density*. Then, label each axis with examples of extremes. For temperature, two extremes would be "cold" and "hot," or if you are more comfortable with numbers use "0" and "1,000,000." And for density, some good labels are "low density" and "high density," or "fluffy" and "dense." Of course, the lower values are normally placed near the origin of the graph.

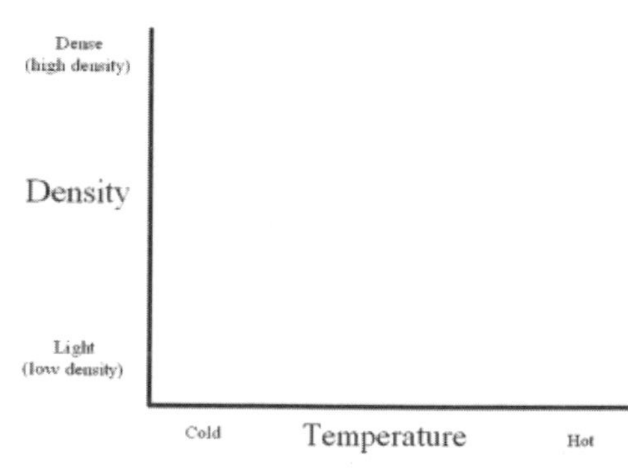

Figure 1.6

At this point all you need to do is ask, "What is the density like for hot materials?" Hot materials, like hot air, are light and fluffy, and they float. So, place a dot where "hot" and "light" cross. "What is the density like for cold materials?" Because cold materials sink, they are dense. Now, place a dot where "cold" and dense" cross.

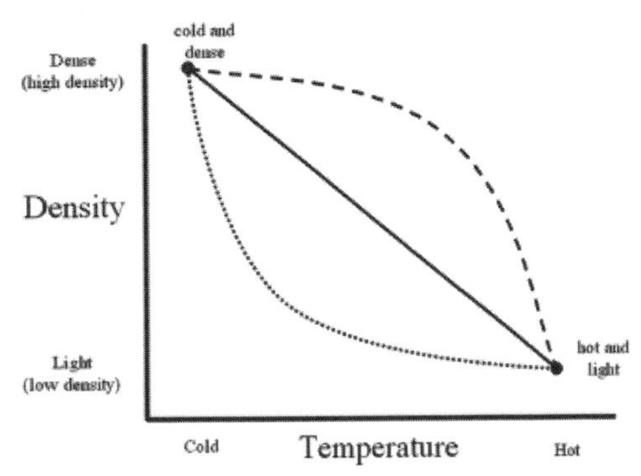

Figure 1.7. Note that even though all three lines begin and end in one place there are different paths that the line can

Once you have two dots, you can connect them. Take note that the path between the dots can be straight or curved, but you will have the general relationship.

Lesson 1–2 Review

1. To make accurate predictions of the future, you need data that is
   a) a random sampling.
   b) a representation of part of a cycle.
   c) a representation of one complete cycle.
   d) made of several cycles.

2. The monthly changes in the phases of the Moon are
   a) random.
   b) cyclic.
   c) not predictable.
   d) highly variable.

3. Tide tables can be printed months in advance because tides
   a) follow a predictable pattern.
   b) are noncyclic events.
   c) happen at the same place at the same time.
   d) are based on the weather patterns.

4. The age of a baby and his or her size are
   a) an inverse relationship.
   b) a direct relationship.
   c) a cyclic relationship.
   d) multiples of each other.

For questions 5 and 6, use Figure 1.8, which shows the changing densities and volumes of a sample of air that is being pressurized.

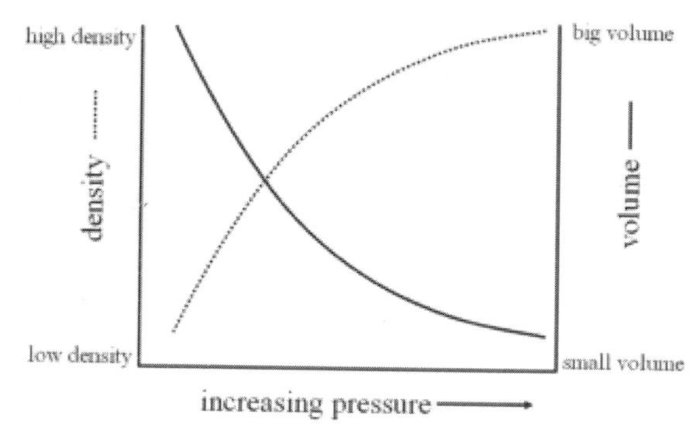

Figure 1.8

5. What kind of relationship does pressure have with density?
   a) inverse b) direct c) cyclic d) no relationship
6. What kind of relationship does pressure have with volume?
   a) inverse b) direct c) cyclic d) no relationship
7. Which graph best shows the relationship between how large an object appears and how close it is to the observer?

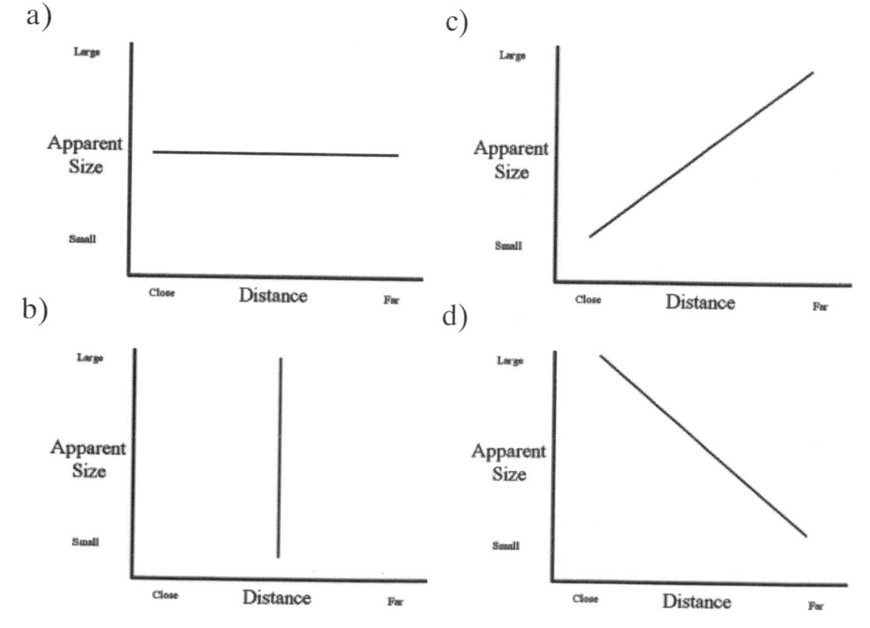

Lesson 1–3: Metric Measurements

When collecting data, such as measurements, there is a wide variety of units that any one thing can be measured with. For example, length can be measured in inches, meters, furlongs, rods, fathoms, or even a "stone's throw." To make sure that all scientific measurements are in the same units, the scientific community has adopted the **Standard International** (SI) system. It is based on the metric system and can be simplified to expressions of grams, meters, and seconds—or combinations of these three main units.

All scientific measurements are taken using the **metric system** (with the exception of American meteorology measurements). There are many advantages to using the metric system. It is used by most of the countries of the world and it allows easy communication between people who speak different languages. It is based on 10s and makes doing math work easy. The system is easier to rebuild from scratch if, for example, your thermometer breaks.

In Earth Science the basic units of the metric system are the **meter** and the **gram**. There are other units used, but these account for the majority of measurements taken for Earth Science calculations.

The metric system is based on water, the most common substance on Earth. Here is an example to illustrate how water can be used to recreate the metric system:

If you shaped 1 gram of liquid water into a cube, it would be a centimeter in each direction (also known as a cubic centimeter or cm^3). One cm^3 is also equal to 1 milliliter (1 mL). If you heat up 1 gram of water with 1 calorie of energy, it will raise the temperature 1°C.

Meter

The meter is the base unit for length. When measuring tiny distances, the meter is too big and is divided into 1,000 pieces called millimeters (*milli* means one thousandth). For distances a little larger, the centimeter is used. It is one hundredth of a meter or 10 millimeters. *Centi* means one hundredth. Just as there is 100 ***cents*** in a dollar, there are 100 ***centi***meters in a meter. For larger distances meters are used, and kilometers are used to measure distances in thousands of meters (*kilo* means 1,000 times).

1 millimeter	1 mm	1,000th of a meter
1 centimeter	1 cm	100th of a meter
1 meter	1 m	Base unit of length
1 kilometer	1 km	1,000 meters

Know Your Sizes

A millimeter is about the thickness of your pinky nail.

A centimeter is about the width of your pinky nail.

A meter is just over a yard and is about the distance from the nose to the fingertips of an adult.

Grams

The **mass** of an object is roughly the measure of how many atoms are within the object. Mass is measured with a scale or balance and is expressed in **grams**. One thousand grams is one kilogram and is about the weight of a pineapple.

1 gram	1 g	Base unit of mass
1 kilogram	1 kg	1,000 grams

Volume

The amount of space an object takes up is **volume**. A good analogy of volume is to take a count of how many sugar cubes fit into an object. A sugar cube is roughly a centimeter on each side—or a cubic centimeter (cm^3). With larger objects, volume is measured in cubic meters (m^3) or even cubic kilometers (km^3).

To measure the volume of an object, there are a few different methods depending on the nature of the object being measured. If the object is a geometric solid such as a block, sphere, or pyramid it is easy enough just to use a formula. For example, a block's volume would be found by using the equation:

Volume = length in centimeters × width in centimeters × height in centimeters

(or $v = l \times w \times h$)

This method will give the results in units of cubic centimeters (cm^3).

To measure the volume of a liquid, you would use a graduated cylinder and the results would be in milliliters (mL). The tricky part of using a graduated cylinder is how exactly to read it. When water is poured into the tube, the water's surface will not be level. It will have a dip in it called a **meniscus**. The volume is read as the value of the bottom of the meniscus. The tubes are designed knowing that the water will dip so don't worry about it not being accurate.

To measure the volume of an irregular solid such as a rock, you would use the **water displacement method**. This measures the volume of the material by how much water is pushed out of a container when the object is submerged. In a lab setting this is done in a graduated cylinder. Fill the cylinder about half-way with water and note the volume of the water. Carefully insert the sample to be measured. When the water rises, note the new water level. The difference between the final level and the starting level is the volume of the rock.

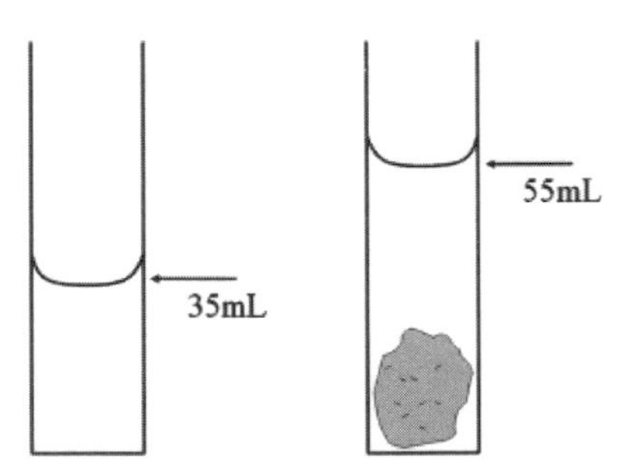

Figure 1.13. In this example, the cylinder with only water has a volume of 35 mL. When the rock is inserted, the new volume is 55 mL. The difference, 20 mL, is the volume that the rock takes up. The rock has a volume of 20 cm³.

One confusing thing about using the water displacement method to measure the volume of a rock is that the units are still cm^3 even though you are using the level of water to find the volume. Let me explain: 1 mL of liquid takes up exactly the same amount of space as 1 cm^3. Therefore, they are interchangeable. Liquids are measured in mLs and solids in cm^3.

Lesson 1–3 Review

1. Measure each line in centimeters (cm) and also in millimeters (mm).
   a) ________________________________
   b) ____
   c) _____________
   d) ____
   e) _________________________

2. a) Draw a line that is 1 cm long.
   b) Now draw a line that is 10 mm long.
   c) How many millimeters are there in 1 cm?
   d) How do you change a measurement from centimeters to millimeters?
   e) How do you change a measurement from millimeters to centimeters?

Lesson 1-4: Density

Density measures how tightly packed an object is. Technically speaking, it is how much mass is in a unit of volume. Some objects can be really dense and feel heavy for their size, whereas other objects are not as dense and feel lighter or fluffier.

Here are two examples of how an object can change its density. Take a piece of aluminum foil and lightly crumple it. Fell how heavy it feels. Now crumple it as hard as you can and feel how heavy it feels now. It will feel heavier. You have just made its volume smaller without changing its mass—it is now denser and feels "heavier" for its size.

For the second example, you start with an empty suitcase. It has a certain volume that it takes up, but because it is empty, it has very little mass. It feels "light" for its size. Now pack it with as many clothes as it will hold. Sit on it to compress to and zip it closed. The suitcase now has a much greater mass without changing its volume—it feels "heavier" for its size.

To calculate density, you divide the object's mass and by its volume. Because mass is measured in grams (g) and volume in cubic centimeters (cm^3), density is measured in grams per cubic centimeter (g/cm^3). The unit g/cm^3 tells us how many grams of material are crammed into each cube the size of a sugar cube.

How to Use the Density Equation

The equation for density is $d = \frac{m}{v}$.

How to use it for finding the volume or mass of a material? The first step is to put the equation into a triangle in this way:

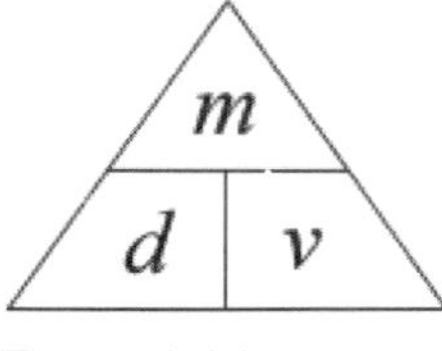

Figure 1.14

In the equation mass, m, is the only variable that is on top of anything else so it goes into the top section of the triangle. The d and v go into the remaining spots and it doesn't matter which letter goes into which section.

In any equation, there is an unknown, which is what you are trying to solve. In the "normal" equation, d is unknown, which you need to find, and m and v are the values that you have. But if you need to solve for m or v, you need to solve a slightly different equation. Here's where the triangle comes into use. Simply use your finger to cover whichever variable you need to solve (you cover it because you don't know what it is). Once you cover the unknown, the remaining variables show you the new equation.

For example, if you need to know what the volume, v, is, cover the v and you are left with mass, m, over density, d. This means that volume is mass over (or divided by) density. If you are left with the two variables next to each other (d and v) then you multiply to find m.

Density: When the Volume Changes

To change the volume of a material, it will either have to be squeezed into a smaller size or expanded so that it takes up more room. But what happens to the density? When a material gets compacted, it packs the material into a smaller space. This might happen if it gets cooler or is put under greater pressure. **When a material gets compacted, the atoms get closer together and the density increases.** When the material expands, it takes up more room—the volume increases. As a result, the atoms get further apart and the density decreases. Materials will expand when they are heated or experience less pressure. **When materials expand, the density decreases; it gets fluffier and has a tendency to float.**

Density: When the Mass Changes

It is an unusual situation to change the mass of an object without affecting its volume, but it can happen. One of the more common examples would be releasing gas from a pressurized canister. When the gas escapes, the mass within the canister decreases but the volume stays the same (the gas within expands to fill the entire canister). **When the mass decreases but the volume stays the same, the density will decrease.** In the same situation, **if more gas is pumped back into the canister, the volume will still stay the same but density will increase.**

Lesson 1–4 Review

1. A sample of quartz has a density of 2.7 g/cm^3. If the sample is cut in half, what will happen to the density of each of the two halves?
   a) The densities of the two halves will be higher than the original.
   b) The densities of the two halves will be lower than the original.
   c) The densities of the two halves will be the same as the original.
   d) There is not enough information.
2. Which of the following is the densest: water in the solid phase, liquid, or gaseous?
   a) solid b) liquid c) gas d) all are the same
3. The reason a hot air balloon rises is because when air is heated
   a) it contracts, making its density lower.
   b) it contracts, making its density higher.
   c) it expands, making its density lower.
   d) it expands, making its density higher.
4. Complete the following chart.

	Density(g/cm^3)	**Mass(g)**	**Volume(cm^3)**
a)		20	5
b)	6		6
c)		18	12
d)	0.25		8
e)	50	15	

5. What is the density of the following sample?

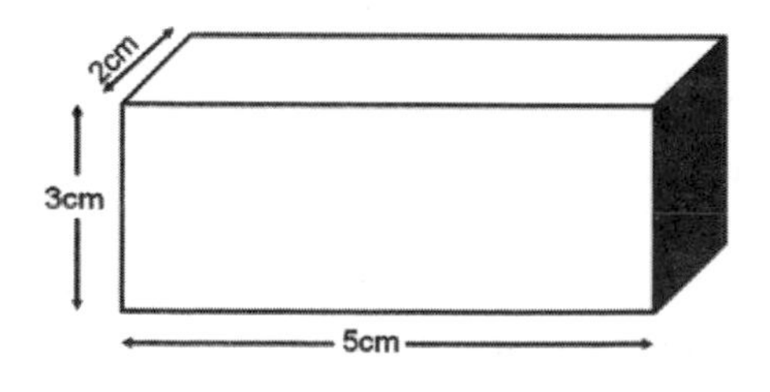

Figure 1.15

Lesson 1–5: Gradient, Rate, and Time

In order to express how quickly values change, we use **rate** and **gradient**. **Rate** tells us how quickly a value changes from minute to minute, while gradient shows how quickly a value changes between two points on a map.

$$\text{rate} = \frac{\text{change in value}}{\text{time}}$$

The "change in value" is simply subtracting one value from another. The order of which number is subtracted from which makes little difference. The only difference would be an answer with a negative sign in front of it, which would just get thrown out.

For example, to measure the weathering rate of a cemetery headstone, find the difference in thickness of the rock (subtract the worn thickness from the original thickness) and then divide by the amount time it took to wear down. If the headstone wore down 18mm in 50 years, then the equation would be:

$$\text{rate} = \frac{18\text{ mm}}{50\text{ years}} = \mathbf{.36\ mm/year}$$

Notice that the units are a combination of the top of the equation and the bottom. The unit "mm/year" is read "millimeters per year" and means that in one year the rock weathered .36 mm.

A good way to think of rate is that it means "speed." It expresses how fast something happens.

Gradient works exactly the same way rate does except instead of time, the values change with the distance on a map. Subtract the two values and divide by the distance between the two values.

$$\text{gradient} = \frac{\text{change in value}}{\text{distance}}$$

As with rate, the units for gradient are a combination of the units on top of the equation and the bottom. This can make some strange combinations such as m/km (meters per kilometer) or m/m (meters per meter). When this happens, there is no need to combine "like terms." ("Meters per kilometer" tells that the slope of a mountain rises so many meters for every kilometer across the map.)

Rate and Time

In Science, it is critical to express ourselves clearly and accurately. One of the most difficult things to express is the difference between rate and time. People often mix up the two. **Rate** tells us how quickly a value changes. Because rate is a speed, we can say the value changes "quickly" or "fast."

Time expresses how much time it takes for an event. Time can never go "fast" or "slow" (except if you are Einstein). But you can have a "short time" or a "long time."

One reason why the two terms are often confused comes from their relationship with each other. Time is a straight measurement of time: t. Rate is a value over time: x/t. In Math class, they would call these two terms "multiplicative inverses," which means that they are exact opposites. If the time gets bigger, the rate gets smaller and vice versa.

For example, if you take a trip to Grandma's house and it takes a long time then you were traveling at a slow rate or speed. On the way back from Grandma's, the same trip takes a shorter amount of time—you were traveling at a faster rate.

> Rate means "speed." When you hear "rate" think of the dial of a speedometer and use words such as *faster* and *slower*.
>
> When you think of time, think of a stopwatch and use phrases such as a *longer time* or a *shorter time*. Never use words such as *quicker* or *slower* to express time.

Lesson 1–5 Review

Exercise A: Gradients

For all problems, show all work and label the answers with the proper units.

1. Write the equation for gradient.
2. Two cities are separated by 300 miles. City X has a temperature of 42°F and city Y has a temperature of 60°F. Calculate the temperature gradient between the two cities.
3. New Earth City has a temperature of 65°F ; Monolith City, 42 miles away, has a temperature of 58°F. Calculate the temperature gradient between the two locations.

4. Chicago is 1,500 miles away from New York. The two areas have a difference in air pressure of 7.5 mb (millibars). What is the pressure gradient?
5. Two cities are 900 miles distant from each other. What is the pressure gradient if the air pressure at one city is 29.21 inches and in the other city it is 29.84 inches?
6. The air pressure inside a tornado can drop 30 mb from the pressure outside it. If an F5 tornado is 1 mile across (5,200 ft), what is the pressure gradient in mb/ft from the outside to the center (half the diameter) of the tornado?

Exercise B: Rates

Complete the following data chart, graph the data (see Figure 1.16 on page 26), and then answer the questions. You will plot two lines on the graph: time and rate. Make your graphs with the time on the left vertical axis and the rate on the right vertical axis.

7. Write the equation for rate of change.

____

Hardness	**Time needed to weather 17 mm of material**	**Rate of weathering (mm/year)**
1	2 years	
2	4 years	
3	8 years	
4	16 years	
5	32 years	
6	64 years	
7	128 years	
8	256 years	

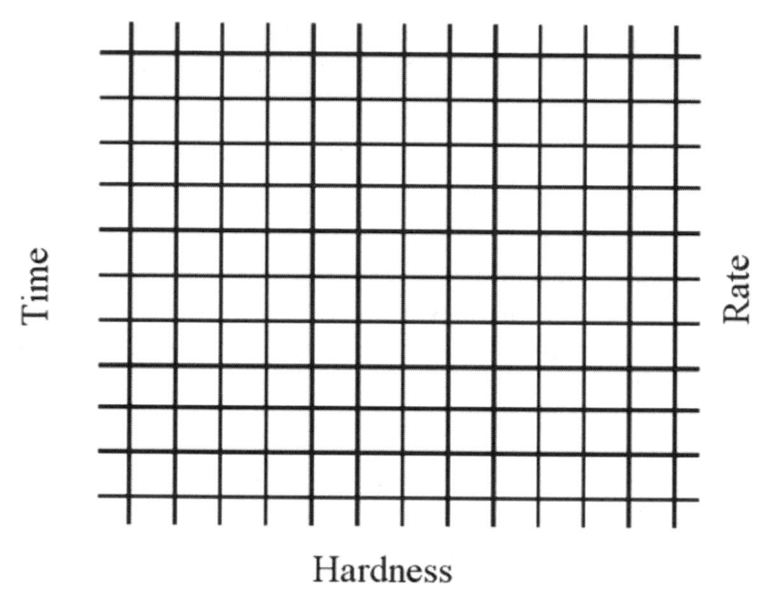

Figure 1.16

8. Which hardness takes the longest *time* to weather?
9. Which hardness has the fastest weathering *rate*?
10. How does the *rate* of weathering change as hardness increases?
11. How does the *time* needed for weathering change as hardness increases?
12. Is the relationship between hardness and weathering time direct or inverse?
13. Is the relationship between hardness and weathering rate direct or inverse?

Lesson 1–6: Percent Deviation

All measurements have some error associated with them. Sometimes it is valuable to know just how large the error is. For example, if you measured a length and you were off by 50 cm, would that be a good measurement? If you just measured a sheet of paper that is about 30 cm long, it would be a horrible measurement. However, if you measured the diameter of Earth (1,275,600,000 cm) and were off by 50 cm, it would be a very good measurement! One way to express error is "percent deviation" or percent error.

Percent deviation takes the value of the error and compares it to the correct value for the measurement.

$$\text{percent deviation} = \frac{\text{difference between correct value and measured value}}{\text{correct value}} \times 100$$

For example: A student measures a rock to have a mass of 54 g while it is actually 50 g.

$$\text{percent deviation} = \frac{54\text{ g} - 50\text{ g}}{50\text{ g}} \times 100$$

$$= \frac{4\text{ g}}{50\text{ g}} \times 100$$

at this point, the grams cancel out

$$= .08 \times 100 = 8\%$$

Lesson 1–6 Review

1. Write the equation for percent deviation.
2. What does "accepted value" mean?
3. A student weighed a rock and found its mass to be 30 g. When checked, she found out that the rock really has a mass of 25 g. What is the percent deviation?
4. A student measured a table to be 150 cm wide. The correct width is 175 cm. What is the percent deviation?
5. Original satellite technology measured the Earth's circumference to be 14,504 km. What is the percent deviation if the actual circumference is 14,564 km?
6. Yor was a Viking. He had to cut wood for a boat. Each piece was to be 22 meters long, but Yor measured them to be 21 meters. What was his percent deviation?

Chapter Exam

1. If a mountain grows at a rate of 400 cm in 100 years, what is its growth rate per year?
2. The lip of Niagara Falls is worn backwards by 900 cm in a month. What is its erosion rate per day (assume a month is 30 days long)?
3. A hiker gains 250 meters in altitude by climbing a mountain from the valley. The peak is 5 km from the valley. What is the gradient of the mountain side?
4. The temperature on one side of a classroom is 27°C; on the other it is 19°C. The distance between the two thermometers is 16 meters. What is the temperature gradient?
5. If the greenhouse effect caused the Earth's average temperature to increase by a total of 3°F between the years 1960 and 1990, what is the rate of temperature change in degrees per year (°F/year)?
6. After a series of earthquakes, a satellite computes that California is 15 cm less wide than it was 3 years ago. What is the rate of change of California's width?
7. As hurricane Felix neared the U.S. coast, air pressure dropped from 996 millibars to 980 millibars in 8 hours. Calculate the rate of air pressure change in mb/hour.
8. Erosion typically wears the Hawaiian volcano Mono Loa down about 20 cm in 5 years. Lava flows build the volcano up about 10 cm every time there is an eruption. Mono Loa erupts about once every 2 years. Is the volcano growing or wearing down?
9. The height of ocean tides over a 3-day period is which type of relationship?

   a) inverse b) direct c) cyclic d) indirect
10. The following statement was made by a student while examining a fossil:

    "It looks like a clam shell, it is 2 cm wide, it lived in shallow-water, and it is embedded in shale."

    Which of the student's statements is an inference?

    a) "it looks like a clam shell" c) "it lived in shallow-water"
    b) "it is 2 cm wide" d) "it is embedded in shale"

11. A distance of 7.81 meters is equal to how many millimeters?
    a) .00781 mm b) .781 mm c) 781 mm d) 7,810 mm

12. Figure 1.17 shows a rock submerged in a graduated cylinder. What is the volume of the rock?
    a) 40 cm^3
    b) 87 cm^3
    c) 47 cm^3
    d) 47 mL

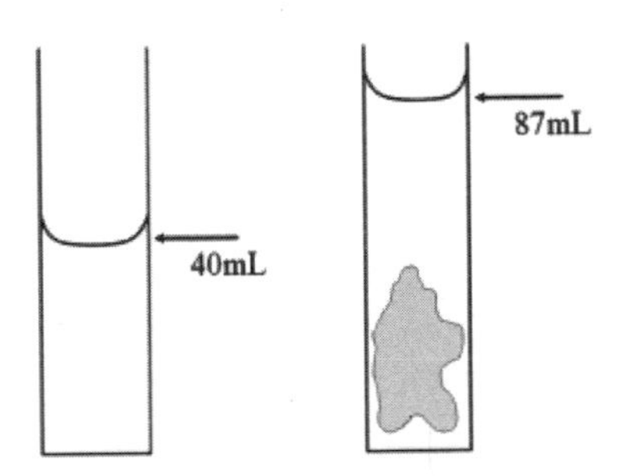

Figure 1.17

13. If a rock has a volume of 51 cm^3 and a mass of 142.8 g, what is its density?

14. What is the volume of a mineral whose density is known to be 5.2 g/cm^3 and has a mass of 33.8 g?

15. If a balloon is filled with room temperature air and is then heated in a microwave so that the heated air expands inside the balloon, what happens to the density of the air within the balloon?
    a) It increases.
    b) It decreases.
    c) Its density is unaffected.
    d) Its density stays the same because the mass will decrease.

16. A granite boulder has a mass of 356.4 kg. A class of students devises a method of indirectly measuring its mass and determines that it has a mass of 361.0 kg. What is the class's percent deviation?

Answer Key

Answers Explained Lesson 1–1

1. **Observation.** The color of the rock can be easily seen.
2. **Observation.** The distance to the Moon can be measured. A measurement is an observation based on a "comparison to a standard."
3. **Inference.** No one saw how the Moon was created. Any theories on its creation are guesses based on the observations we can make.
4. **Observation.** Rain can be seen, felt, smelled, and heard.
5. **Inference.** Predictions of the future based on current and past observations are inferences.

6. **Inference**. No one saw the glacier eroding the landscape. It can only be guessed based on observations such as the types of scratches and sediments in the area.
7. **Inference**. The primitive humans didn't leave a record behind telling that they killed all of the mammoths. The mammoths could have also gone extinct from climate change or disease.

Answers Explained Lesson 1–2

1. **A** is incorrect because a random sampling will not always give enough information to notice a pattern.

   B is incorrect because part of a cycle is not enough information on which to base a prediction.

   C is incorrect because there may be variations from one cycle to the next causing an accurate prediction to be off.

   D is **correct** because several cycles give a better chance to spot any variations from one cycle to another.
2. **A** is incorrect.

   B is **correct** the Moon's phases follow a highly regular and repeating pattern.

   C and **D** are incorrect because the Moon's phases are predictable.
3. **A** is **correct** because the tides follow the regular phases of the Moon, which are predictable.

   B is incorrect because tides are very cyclic. They go through a cycle twice a day (in most locations) with an overall pattern that repeats every month.

   C is incorrect because different tides happen at different places at different times but the same tides happen at regular intervals in one location.

   D is incorrect because tides are based on the phases of the Moon.
4. **A** is incorrect. If it were inverse, the baby would get smaller as he or she got older.

   B is **correct**. As the baby gets older, he or she gets larger. They both go up so it is a direct relationship.

   C is incorrect because this would mean that the baby gets bigger and then smaller, over and over.

   D is incorrect because the relationship between age and height is a complicated relationship—not simple multiples of each other.
5. **A** is incorrect because this would mean that the more an object is squeezed, the fluffier it gets.

   B is **correct** because as the pressure increases the object gets more compacted, making it denser.

   C is incorrect because this would mean that the object will get denser, then lighter, then denser as the pressure increases.

   D is incorrect because pressure affects the density by squeezing the material.

6. **A** is **correct** because as pressure increases, the volume get squeezed smaller.

   B is incorrect because this would mean that as the pressure gets higher, the volume would also get larger. In other words, when the object gets squeezed, it gets bigger—which does not happen.

   C is incorrect because a cyclic relationship means that the volume would keep getting bigger and smaller when squeezed.

   D is incorrect because pressure affects the volume—there is a relationship.

7. **A** is incorrect because graph A shows that the apparent size of the object stays the same regardless of its distance. Objects will look smaller as they get further away.

   B is incorrect because graph B shows that at a certain distance, let's say 20 feet, the object will appear to be all sizes which is not possible.

   C is incorrect because graph C shows that as the object gets further away it appears larger.

   D is **correct** because graph D shows that as the object gets further away, it appears to get smaller.

Answers Explained Lesson 1–3

1. a) **6.0 cm; 60 mm**

   b) **8.5 cm; 85 mm**

   c) **2.5 cm; 25 mm**

   d) **10.1 cm; 101 mm**

   e) **4.8 cm; 48 mm**

2. a) and b) Both lines should be the same length: ———

   c) **There are 10 millimeters in one centimeter**. Each tiny line between the centimeter marks is a millimeter. You might only count 9 small mm lines between cm marks because the 10th mm line is the next centimeter.

   d) To change from centimeters to millimeters, you will need to multiply by 10. Because millimeters are smaller, you will need more of them.

   e) To change from millimeters to centimeters, you will need to divide by 10. Because centimeters are larger, you will need less of them.

Answers Explained Lesson 1–4

1. **C** is **correct**. The density will stay the same. If you cut a sample in half, you remove half of the mass **as well as** half of the volume. In Math class, your teacher will explain it as "if you do the same thing to the top and bottom of a fraction (mass and volume), then you don't change the value of the fraction." For example: If the mass is 10 g and the volume is 10 cm^3, then the density will be 1 g/cm^3. Cut it in half and you have 5 g and 5 cm^3—still a density of 1 g/cm^3.

2. **A** is incorrect because solid water (ice) floats on liquid water, which means that the solid is less dense.

   B is **correct** because the liquid phase of water is densest, which is why it is found at the bottom of a mixture of solid liquid and gas. NOTE: Water is very unusual in this fashion. Most materials are densest in their solid phase.

   C is incorrect because water vapor is very light, which is why water vapor floats into the air to eventually condense into clouds.

   D is incorrect because if all phases have the same densities then they would all float at the same levels.

3. **A** is incorrect. Air contracts when it is cooled. When a material contracts, its volume decreases which makes its density go up—it gets "heavier."

   B is incorrect because this would happen if the air is cooled.

   C is **correct** because when air is heated it expands increasing its volume. A larger volume would lower its density making it fluffier and it will want to float.

   D is incorrect because when a material expands its density gets lower.

4. Answers are calculated below the chart.

	Density(g/cm^3)	Mass(g)	Volume(cm^3)
a	**4**	20	5
b	6	**36**	6
c	**1.5**	18	12
d	0.25	**2**	8
e	50	15	**.3**

a) $20\ g \div 5\ cm^3 = 4\ g/cm^3$

b) Use the triangle to change the equation to $m = d \times v$

$6\ g/cm^3 \times 6\ cm^3 = 36\ g$

c) $18\ g \div 12\ cm^3 = 1.5\ g/cm^3$

d) Use the triangle to change the equation to $m = d \times v$

$0.25\ g/cm^3 \times 8\ cm^3 = 2\ g$

e) Use the triangle to change the equation to $v = m \div d$

$15\ g \div 50\ g/cm^3 = .3\ cm^3$

5. The volume of the block is length × width × height ($l \times w \times h$)

   $5\ cm \times 2\ cm \times 3\ cm = 30\ cm^3$

   Because the mass is 20 g the density will be $20\ g \div 30\ cm^3 =$ **$.66\ g/cm^3$**

Answers Explained Lesson 1–5

Exercise A

1. $\text{gradient} = \dfrac{\text{change in value}}{\text{distance}}$

2. The change in value is 60°F – 42°F, which equals 18°F. The distance is 300 miles.

   $\dfrac{18\ °\text{F}}{300\ \text{miles}} = .06\ °\text{F/mi}$

3. The change in value is 65°F – 58°F, which equals 7°F. The distance is 42 miles.

   $\dfrac{7\ °\text{F}}{42\ \text{miles}} = .17\ °\text{F/mi}$

4. $\dfrac{7.5\ \text{mb}}{1500\ \text{mi}} = .005\ \text{mb/mi}$

5. The change in value is 29.84 in. – 29.21 in, which equals .63 in. The distance is 900 miles.

   $\dfrac{.63\ \text{in}}{900\ \text{miles}} = .0007\ \text{in/mi}$

6. The distance in this problem is 2,600 ft (if the tornado is 5,200 feet across then it is 2,600 feet to the center).

   $\dfrac{30\ \text{mb}}{2{,}600\ \text{ft}} = .012\ \text{mb/ft}$

Exercise B—Rates:

7. $\text{rate} = \dfrac{\text{change in values}}{\text{time}}$

Hardness	Time needed to weather 17 mm of material	Rate of weathering (mm/year)
1	2 years	**8.5**
2	4 years	**4.25**
3	8 years	**2.15**
4	16 years	**1.06**
5	32 years	**.53**
6	64 years	**.27**
7	128 years	**.13**
8	256 years	**.07**

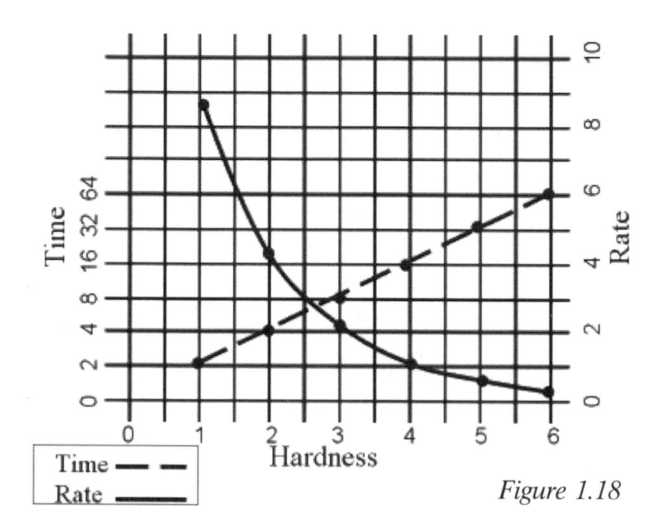

Figure 1.18

8. A hardness of 8 takes the longest with 256 years needed to weather 17 mm.
9. A hardness of 1 has the highest weathering rate of 8.5 mm/year.
10. The rate of weathering decreases as the hardness increases. The harder something is, the slower it will wear away.
11. As hardness increases, the time needed to weather will also increase. The harder a material, the longer it will take to wear it away.
12. The relationship between hardness and weathering time is *direct* because the harder a material, the longer it takes to weather—they both increase.
13. The relationship between hardness and weathering rate is *inverse* because as the hardness increases, the weathering rate decreases—as one goes up the other goes down.

Answers Explained Lesson 1–6

1. percent deviation

$$= \frac{\text{difference between correct value and measured value}}{\text{correct value}} \times 100$$

2. Accepted value means the "correct" value or the true value.

3. percent deviation $= \frac{30\text{ g} - 25\text{ g}}{25\text{ g}} \times 100 = \frac{5\text{ g}}{25\text{ g}} \times 100 = .2 \times 100 = 20\ \%$

4. percent deviation $= \frac{175\text{ cm} - 150\text{ cm}}{175\text{ cm}} \times 100$

$$= \frac{25\text{ cm}}{175\text{ cm}} \times 100 = .143 \times 100 = 14.3\%$$

5. $$\text{percent deviation} = \frac{14{,}564\text{ km} - 14{,}504\text{ km}}{14{,}564\text{ km}} \times 100$$
$$= \frac{60\text{ km}}{14{,}564\text{ km}} \times 100 = .004 \times 100 = 0.4\%$$

6. $$\text{percent deviation} = \frac{22\text{ m} - 21\text{ m}}{21\text{ m}} \times 100 = \frac{1\text{ m}}{21\text{ m}} \times 100 = .048 \times 100 = 4.8\%$$

Answers Explained Chapter 1 Exam

1. 400 cm ÷ 100 years = **4 cm/year**
2. 900 cm ÷ 30 days = **30 cm/year**
3. 250 m ÷ 5 km = **50 m/km**

   Note that in this case the proper units are m/km. You do not need to convert km into m to make the units the same. This gradient means "you gain 50 meter for every km walked."
4. $$\frac{27\,°\text{C} - 19\,°\text{C}}{16\text{ m}} = \frac{8°\text{C}}{16\text{ m}} = \mathbf{.5°C/m}$$
5. 3 °F ÷ 30 years = **0.1°F/year**
6. 15 cm ÷ 3 years = **5 cm/year**
7. $$\frac{996\text{ mb} - 980\text{ mb}}{8\text{ hours}} = \frac{16\text{ mb}}{8\text{ hrs}} = \mathbf{2\ mb/hr}$$
8. Erosion is happening at a rate of:

   20 cm ÷ 5years = **4 cm/year**

   Growth is happening at a rate of:

   10 cm ÷ 2 years = 5 cm/year

   Therefore the volcano is getting larger by 1 cm/year.
9. **A** is incorrect because this would mean that as time went on (increased) the tides would get lower (decrease).

   B is incorrect because this would mean that as time went on (increased) the tides would also get higher (increase).

   C is **correct** because the tides rise and fall repeatedly throughout the day.

   D is incorrect because this would mean that as time went on (increased) the tides would get lower (decrease).
10. **A** is incorrect because this is an observation. The description, whether it is accurate or not, still describes what it looks like.

    B is incorrect because measurements are observations.

C is **correct** because the sample is a fossil and no one saw exactly where it lived. Any guesses about its environment must be an inference.

D is incorrect because this is an identification of the type of rock in which the fossil was found—an observation.

11. **D is correct**. To convert meters to millimeters you need to multiply. Because millimeters are smaller than meters, you will need more of them. There are 10 mm in each cm and there are 100 cm in a meter. To convert, multiply 7.81 m and multiply by 1000. 7.81 × 1000 = 7,810 mm.

12. **A** is incorrect because the starting volume of water was 40 cm^3.

    B is incorrect because the volume of the water and the rock came to a total of 87 cm^3.

    C is **correct** because 87 mL – 40 mL = 47 mL. Because you are measuring the volume of a solid you convert the mLs into cm^3s.

    D is incorrect because the volume of the displaced water is 47 mL, but because you are measuring the volume of a solid, it should be converted into cm^3s.

13. **2.8 g/cm³**

$$d = \frac{m}{v}$$

$$d = \frac{142.8 \text{ g}}{51 \text{ cm}^3} = 2.8 \text{ g/cm}^3$$

14. **6.5 cm³**

Change the equation from $d = \frac{m}{v}$ to $v = \frac{m}{d}$

$$v = \frac{33.8 \text{ g}}{5.2 \text{ g/cm}^3} = 6.5 \text{ cm}^3$$

15. **B** is **Correct**. When a material is heated and its volume increases but its mass remains the same, the density will drop. For example: If the original mass and volume are 10 g and 10 cm^3, the density will be 1 g/cm^3. When the volume increases to 20 cm^3, the density will drop to .5 g/cm^3.

16. $$\text{percent deviation} = \frac{\text{difference from accepted value}}{\text{accepted value}} \times 100$$

$$\text{percent deviation} = \frac{361.0 \text{ kg} - 356.4 \text{ kg}}{356.4 \text{ kg}} \times 100 = \frac{4.6 \text{ kg}}{356.4 \text{ kg}} \times 100$$

$$= 0.0129 \times 100 = 1.29\%$$

Note that the class has the incorrect value because the question states that the rock is 356.4 kg and also asks "what is the class's percent deviation."

1 2 3 4 5 6 7 8 9 10

Dimensions of the Earth

Lesson 2–1: Shape of the Earth

Contrary to popular belief, Christopher Columbus did not discover that the world was round! It was known for centuries before Columbus came along. Some basic observations lead to the implication of Earth's true shape: round.

Ships Sailing Over the Horizon

One of the earliest observations came from watching sailing ships go "over the horizon." As a ship sails away from shore, it appears to sink slowly once it gets many miles out to sea. What is actually happening is the boat is going over the curvature of the Earth. To the sailors on the boat, they will see the land appear to sink as well. However, when returning from a voyage, the ship will appear to emerge out of the water in the opposite fashion, and everything is found in order when the sailors safely return home.

Eclipses

Another ancient observation of Earth's curvature is the shape of its shadow on the Moon during an eclipse. Ancient astronomers realized that the shadow was caused by the Sun, Earth, and Moon aligning in a straight line, with Earth in the middle casting its shadow upon the Moon. The edge of the shadow was always curved despite the time of day of the eclipse. This meant that the Earth casts a rounded shadow from every angle. The only object that appears round from all directions is a sphere.

Time Zones

The fact that we have different time zones on the planet is caused by the curvature of the Earth. If the world was flat, the Sun would rise everywhere at the same time and we would all set our clocks the same. Because the world is round, the Sun rises at different times as you travel east or west. The planet is divided into 15° bands in which all clocks are set to the same hour.

Angles to Stars

The angle between the horizon and Polaris (the North Star) changes as you travel north or south. As you get closer to the North Pole, Polaris gets higher in the sky and gets closer to a 90° angle to the horizon. As you travel south, Polaris gets lower in the sky. If you travel too far south (past the equator), Polaris sinks below the horizon just as boats as they sail away.

Probably the best evidence of Earth's true shape comes from direct observation. Astronauts have actually seen the curvature of Earth and have taken pictures from space. With these pictures and highly accurate measurements taken from space, the exact shape of Earth can be measured.

Evidence that the Earth is round:

1. Ships sailing over the horizon.
2. The Earth's shadow on the Moon during an eclipse is always curved.
3. Time zones.
4. The angle to Polaris changes as you travel north or south.
5. Photos from space (the best proof).

The Earth is very close to a perfectly round sphere. However, due to its spinning once a day around its axis, the Earth is slightly larger at the equator. As the equator gets "flung" out into space a little, the poles get pulled inward. This flattens the Earth at the poles and bulges at the equator, making the true shape of Earth a slightly, oblate spheroid (*oblate* means squashed and *spheroid* means ball).

The difference between the length of the diameter going from pole-to-pole (up-and-down) and the diameter through the equator is not much—only 42 km. For a planet that is 12,756 km across, it is a miniscule difference—so small that it cannot be seen to be imperfect at all. To put it in perspective, if you drew two lines next to each other, one 12,756 km and the other 12,714 km and looked at them from space, you would not be able to see the difference of 42 km in the length of the lines.

Using a ratio between the length of the diameter going from pole-to-pole (up-and-down) and the diameter through the equator (side-to-side) the roundness can be measured.

The equation for calculating the roundness ratio of a round object is:

$$\text{roundness ratio} = \frac{\text{polar diameter}}{\text{equatorial diameter}}$$

For the Earth:

$$\text{roundness ratio} = \frac{12{,}714\ \cancel{\text{km}}}{12{,}756\ \cancel{\text{km}}}$$

(The kms cancel out.)

$$\text{roundness ratio} = \frac{12{,}714}{12{,}756} = .9967$$

The roundness ratio for Earth is .9967, which is 99.67% of a perfect sphere!

Another indication of Earth's true shape comes from gravity measurements. Because the equatorial diameter is slightly larger than the polar diameter, objects at the equator are slightly further away from Earth's center. Because they are further away from the center of the Earth, gravity pulls on them with less force and they are lighter. The difference is very slight but can be measured with sensitive scales.

Similar to Earth's slightly imperfect roundness, the sizes of the bumps on Earth are so small, compared to the size of the planet, that the Earth appears to be perfectly smooth from space. The largest bump in the world, Mt. Everest, is a mere 8.8 km compared to the 12,756 km size of our world.

A good model of the planet would be a smooth pool ball. On paper, the Earth's shape is indistinguishable from a perfect circle.

To sum up the Earth's shape and smoothness: The Earth is not perfectly round nor perfectly smooth. It is a slightly (slightly, slightly) oblate spheroid with mountains and valleys. However, as seen from space, the Earth is so close to perfect (99.7%) that we would not be able to perceive its imperfections.

Lesson 2–1 Review

1. Write the equation for calculating the roundness ratio of a planet.
2. Jupiter has a visibly large oblateness. It has a polar diameter of 133,708 kilometers and an equatorial diameter of 142,984 kilometers. Calculate Jupiter's roundness ratio.
3. Calculate the Moon's roundness ratio. The polar diameter is 3,472 km and the equatorial diameter is 3,476.2 km.
4. What is Mars's roundness if its polar diameter is 6,750 km and its equatorial diameter is 6,794 km?
5. The Death Star from Star Wars has a polar diameter on a computer screen of 1,682 pixels; the equatorial diameter is 1,686 pixels. What is its roundness ratio?

Lesson 2–2: The Size of the Earth

The Story of Eratosthenes

Eratosthenes was a Greek geographer and mathematician (and also a librarian!) who lived in Egypt over 2,000 years ago and is credited as being the first person to accurately measure the size of the Earth. He heard a story of a famous well in a distant city that, on the longest day of the year, at exactly 12 o'clock noon, the Sun would shine directly down the well in the middle of the city. At exactly the same time Eratosthenes looked down the well in his own city. The Sun wasn't shining straight down the well in his town while, at the same time, the Sun was shining straight down the well in the other town.

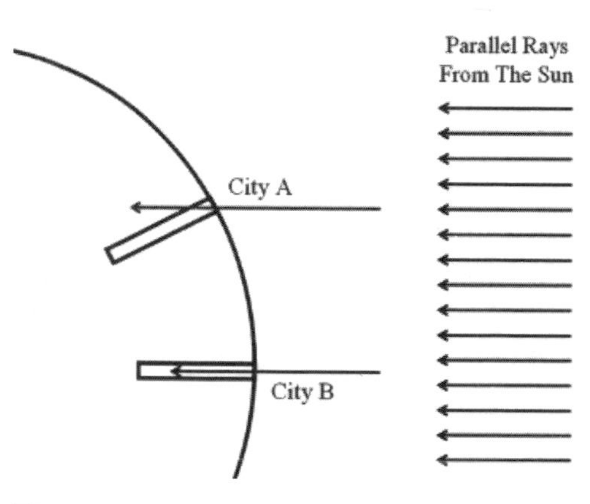

Figure 2.1 Two Wells

Eratosthenes concluded that the simplest explanation for this was that the surface of the Earth was curved and that the famous well was on the spot of the curve directly under the Sun, whereas his home town was off to the side a little.

Eratosthenes developed an equation to calculate the entire size of the Earth using the little information he had available.

$$\frac{A}{360°} = \frac{D}{C}$$

A is the angle between the two cities as measured from the core.

D is the distance between the two cities.

C is the circumference of Earth.

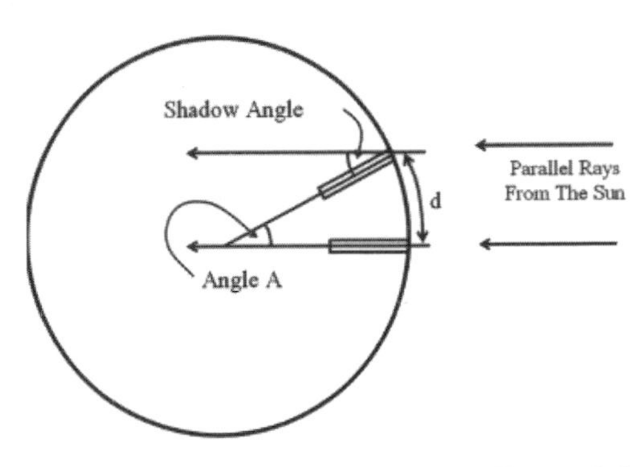

Figure 2.2

Even though Eratosthenes couldn't get to the center of the Earth to measure A, he could calculate it. In geometry, angle A is the same as the angle the shadow in his well makes with the well wall—otherwise known as "alternate interior angles."

Think of this equation as a proportion:
Angle A is to 360° as distance D is to the whole circle.

Because C is the unknown, the equation can be rearranged to:

$$C = \frac{360°}{A} \times D$$

One way of managing this equation is to think of pizza! If you know the angle of one slice of pizza, you can tell how many slices will fit in a pie—which is the fraction part of the equation. Then multiply it by D, which, in pizza, is the length of crust for one slice.

Although Eratosthenes had a perfect system for measuring the circumference of Earth, there were some limitations that caused his final result to be off a little. For example, the two cities needed to be exactly north and south of each other and the accuracy of his measurement of the shadow angle was off a little.

Lesson 2–2 Review

1. Write Eratosthenes' equation for finding the circumference of the Earth.

Calculate the circumference of the following circles.

2.

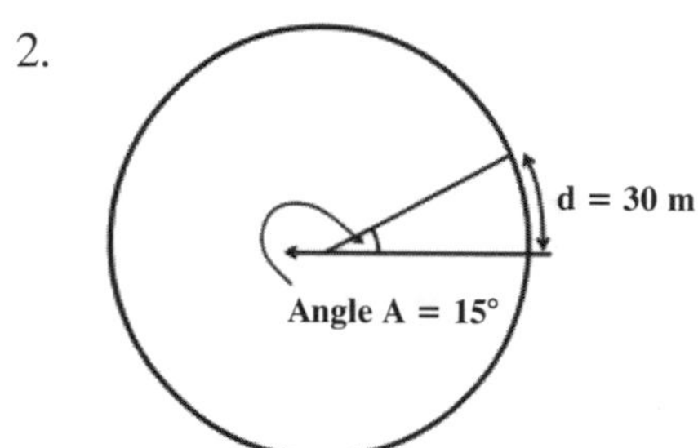

Figure 2.3

3.

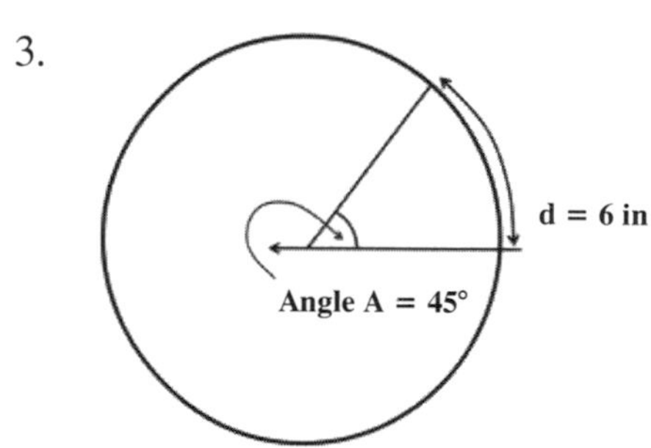

Figure 2.4

4.

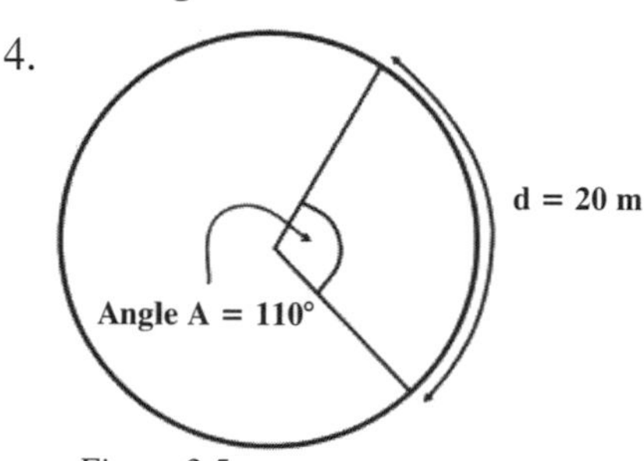

Figure 2.5

5.

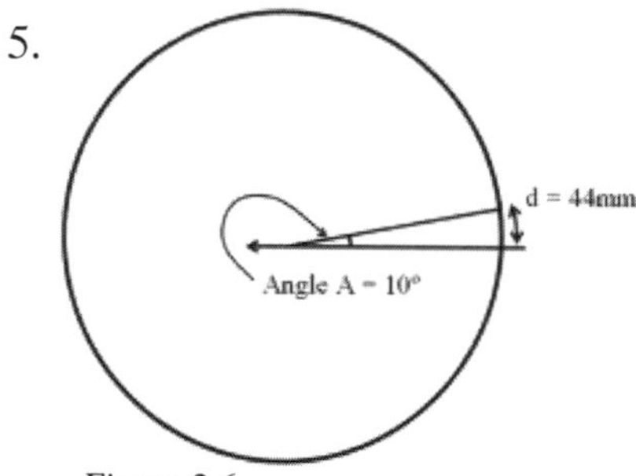

Figure 2.6

Lesson 2–3: Parts of the Earth

The Earth's structure is layered. Each layer is called a "sphere," because it wraps around the ball of Earth.

The **lithosphere** (*lith* means "rock") is the rock layer of the Earth. The lithosphere goes from the surface of Earth downward and includes the **crust** and **upper mantle**. The crust is made of two distinct types of material: continental crust and oceanic crust. The oceanic crust is dark, dense basalt and, as the name implies, lies at the bottom of the oceans. The continental crust is primarily light-colored granite, is thicker, and has a lower density than the oceanic crust. As a result, the continental rocks stick up higher than the oceanic and the dense oceanic rocks sink lower into Earth's surface. Even without large bodies of water covering three quarters of the planet, there would be a noticeable difference between the two regions. When the water began to cover the surface, the water filled in the lowest areas, which happens to be where the oceanic crust covers the surface. Beneath the lithosphere, the interior of Earth is layered (which will be discussed in more detail in Chapter 4).

The **hydrosphere** is the water layer of the Earth. Primarily, this includes the oceans, but it also includes lakes, rivers, groundwater, ice caps, and atmospheric water.

The **atmosphere** is the air layer of the Earth. The atmosphere extends upwards from the surface into space. There is no definable upper limit to the atmosphere because it continues to fade as you go higher and higher. Different sources will give different heights of the "top" of the atmosphere depending on what definition is used to draw the line.

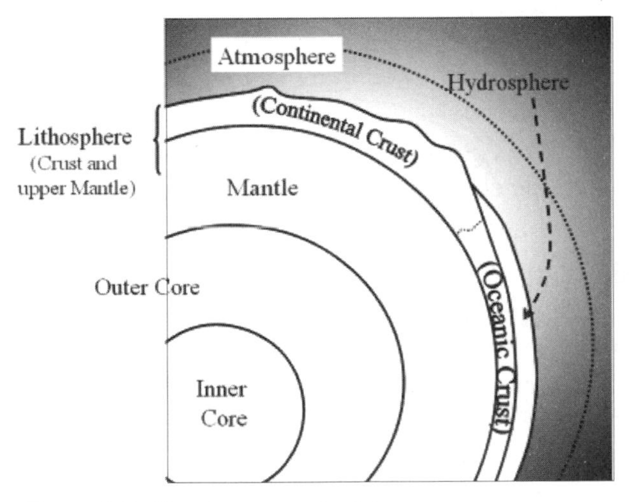

Figure 2.7 (Not Drawn to Scale)

Note on scale: The atmosphere and hydrosphere are so thin compared to the entire size of the Earth that, if drawn as seen from space, the ink line would be wider than each of the two layers would be if drawn to scale.

Lesson 2–3 Review

1. Name the major layers of the Earth from the top to the bottom.
2. What are the parts of the lithosphere?
3. Which layer can be found between the crust and the atmosphere?
   a) mantle b) hydrosphere c) oceanic crust d) lithosphere

Lesson 2-4: Positions on the Earth

Latitude and Longitude

To be able to describe exactly where you are on the Earth, a system of coordinates has been invented using latitude and longitude. Basically, the system divides the planet into a grid and reads very much like a graph with the origin (0,0) in the center.

At the equator, two lines of latitude or longitude 1° apart are separated by roughly 70 miles. This is too crude of a grid to accurately navigate so the degree was separated into smaller units called "minutes." A degree is divided into 60 **minutes** (the symbol is ′ the same as the symbol for foot). A minute of arc is also a little big for accurate measurement so it is divided into 60 **seconds** (the symbol is ″ the same as inches). Using degrees, minutes, and seconds, any spot on earth could be plotted to within 100 feet or so. If you need to be more accurate than a second, the system then starts to use decimal places.

For example, the coordinates for Devil's Tower in Wyoming, the central feature from the classic movie *Close Encounters of the Third Kind* are:

44° 35′ 30.8″N and 104° 42′ 52.7″W

and is pronounced "44 degrees, 35 minutes, 30 point 8 seconds North and 104 degrees, 42 minutes, 52 point 7 degrees West."

Latitude is your angle north or south of the equator as measured from the core of the Earth. In the Northern Hemisphere, we have an easier time measuring our Latitude with the aide of Polaris (the North Star). Polaris is lined up almost exactly with the Earth's axis of rotation. As a result, if you were standing at the North Pole, Polaris would be directly above your head, or, in other words, 90° above the horizon. This works out perfectly because the latitude of the North Pole is 90°N (90° away from the equator). **In the Northern Hemisphere, the angle to Polaris equals your latitude.**

What is GPS?

GPS stands for "Global Positioning System." It is a network of satellites surrounding Earth that are used for finding positions with high accuracy. GPS receivers are devices that compare signals from several satellites in order to get an exact location. They have an accuracy of being able to locate a person within feet or inches.

GPS systems are becoming more common and can be found in cars, camping equipment, hand-held units and of course plane and boat navigation equipment. There are even systems available that are being used for child safety!

Latitude is a natural system that comes automatically with our spinning planet. Because the Earth spins, it has an axis of rotation which gives us a North and South Pole and therefore an equator. As a result, the latitude system comes with preset landmarks—"starting points," if you will. Longitude, however, does not have natural "starting points." The selection of where "zero" is located is a completely arbitrary decision. It was selected to be Greenwich, England, because a "Longitude Committee" decided to put it there. It is now called the **Prime Meridian** ("First Meridian") and it is the starting point for longitude measurements.

All longitude lines, called **meridians**, run from the North Pole, down one side of the Earth, and then to the South Pole. Your **longitude** is how many degrees you are away from the Prime Meridian. Finding the longitude of a city would be like standing at the North Pole, pointing one hand at England and the other hand at the location you want. The angle between your arms would be the longitude. Everything to the left of England (as seen from the North Pole) up to the 180° line of longitude is the Eastern Hemisphere and everything to the right is the Western. You can travel up to 180° away from the Prime Meridian and then you start getting closer on the other side.

One thing that confuses some people about traveling across the globe and then passing the 180° line is what hemisphere you are in (or more commonly the problem is which letter, E or W, goes after a coordinate). If you travel to the west and reach the United States, you are in the Western Hemisphere. If you continue traveling until you pass the 180° line, you will now be in the Eastern Hemisphere but still traveling west. Think of it this way: you can travel all around the world heading always west. When you are in China (the Eastern Hemisphere) you are still allowed to travel west!

How to Measure Your Coordinates

To measure your latitude, measure the angle to Polaris from the horizon. Of course, this will be your degrees north. (In the Southern Hemisphere, there is no "South Star," but the spot where one should be can be easily found by using a constellation called the "Southern Cross.")

To measure your longitude, calculate the time difference between you and Greenwich, England. Because the Earth rotates at 15° per hour, multiply your time difference by 15° to get your longitude. If your time is earlier than England's, you are west, and if you are later, then you are east.

To Sum Up Latitude and Longitude

- Coordinates are divided into degrees, minutes, and seconds.
- Latitude lines run side-to-side and measure how far you are north or south from the equator.
- The angle to the North Star (Polaris) is equal to your latitude (only works north of the Equator).
- Longitude lines run up-and-down and measure how far you are from England, either east or west.

Longitude and Time

The Earth spins once every 24 hours. There's no big surprise there! But in one hour, the earth spins only 1/24 of a full circle (360°), or 15°. This means that if you watched a star at night, its position will shift by 15° every hour until it sets in the west. The Sun, which is a star as well, also moves across the sky at 15° per hour. If you instantly traveled west across the Earth 15° of longitude, it would take the Sun one hour to "catch up" and reach the same position in the sky.

If everyone set their clocks so that sunrise happened at exactly 6 a.m., then everybody's clocks would read different times depending on how far east or west they were. To make clocks more agreeable, it has been decided that everyone who lives within a 15° zone will set their clocks to the same time. In the middle of the zone, the clocks would agree perfectly with sunrise (not accounting for seasonal differences). People to the far edges of the 15° zone would have their clocks as much as a half-hour (plus or minus depending if you were east or west of the center) off of the solar time. These 15° zones are the **time zones** and they allow everyone within an area to arrive at work at the same time—otherwise, people who live to the west would always be late to work!

Time zones are actually *approximately* 15° across. The lines zig and zag to avoid running through populated areas. Imagine the confusion if half of a major city was at 7 p.m. while the other half was at 8 p.m.! There are even a few places on Earth where they have "half time zones" where the zones are 7.5° wide. These places will have the minute hands of their clocks off by 30 from the rest of the world.

Lesson 2–4 Review

1. What is the largest internal angle of latitude that can be measured from the equator?
   a) 180° b) 45° c) 90° d) 360°

2. What is the latitude of the North Pole?
   a) 90°N b) 0° c) 90°E d) 180°

3. What is the latitude for the observer in Figure 2.8?
   a) 90° c) 30°
   b) 60° d) 0°

4. What is the maximum number of degrees of longitude?
   a) 45° c) 360°
   b) 90° d) 180°

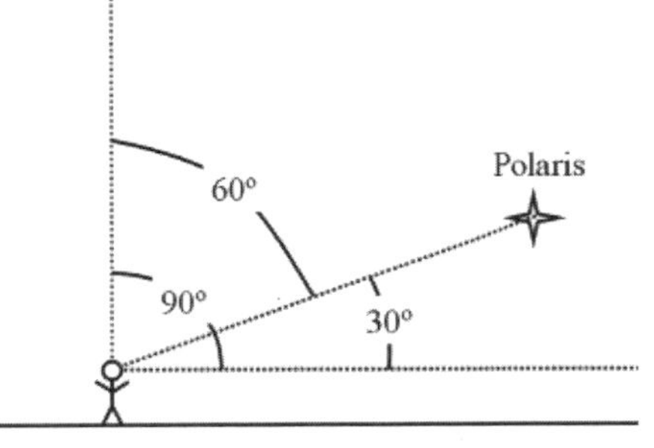

Figure 2.8

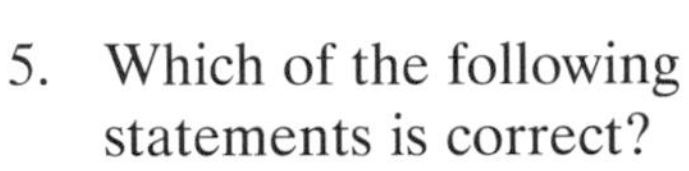

5. Which of the following statements is correct?
   a) Longitude is the angular distance measured east or west of the Prime Meridian.
   b) Lines of latitude, called meridians, never touch.
   c) The Prime Meridian is a full circle that passes through the North Pole and the South Pole.
   d) Latitude is the angular distance measured east or west of the equator.

Lesson 2–5: Use of Contour Maps

Isolines

When discussing maps, a **field** is a region where a value can be measured at any point. Some examples of fields are temperature, light intensity, elevation, pollution concentration, and noise levels. Each of these values can be measured at any point in an area. An **isoline** is used on a map to connect points that have the same value. (*Iso-* means "the same," as in isosceles: a triangle with two equal sides.)

To draw isolines, there are a few simple rules:

1. Isolines connect points of equal value. *Every* spot on the line has the same value—even if nobody stood at that spot and measured the value.
2. Isolines are gentle, curving lines—no sharp corners. Values change gradually. Sharp turns on an isoline map means that the value changed very suddenly.
3. Isolines are always closed curves even though the map might only show part of it. For example, the 80° temperature line will typically wind a path across the United States and stop at each coast. However, that line will continue across the oceans and around the world to eventually complete a loop. A map of the United States is just too small to show the entire loop.
4. Isolines *never* cross—this would mean that the point where the two lines cross has two different values. That would be like having one spot with two temperatures!
5. Isolines usually are parallel. Not parallel in the math sense—rather, they have a parallel trend. In other words, they tend to follow each other. They curve and dip together.

The interval of an isoline map is the difference in value from one line to the next. On a contour map, the **contour interval** represents the amount that the elevation changes from one line to the next. The interval is chosen depending on how much the elevation changes on the map—which is referred to as the **relief**. Even though there is no set amount for the contour interval, there is one thing that helps count the value of the lines: Every fifth contour line is bold and is labeled somewhere on the line.

Another help on contour maps are benchmarks. A **benchmark** is a spot on the map where a person actually stood and accurately measured the exact elevation. A benchmark is shown with an X marking the position and the value of the elevation. If a benchmark is made at the peak of a hill, there will be a triangle shown instead of the X along with the elevation of the summit.

How to Draw Isolines

1. Start by choosing a value that is in the middle near the middle. In other words, you want a value that is both away from the edges of the map and in the middle of the range of values on

the map. This will give you the best chance of starting with a complete line. All the other lines will be variations of this line because they run parallel to each other.

2. Darken or highlight all the data points with your pencil. If there aren't enough data points, you will need to "create" a few of your own. If there are not enough dots to connect to make a line, add some more data points. The new points will be in the location where the value belongs except that the value was not written on the map. For example, if you are looking for all 500 meter values, there will be one between two points that are 450 m and 525 m. For accuracy, the data point will be closer to the 525 m point.
3. Connect the dots of the same value with a smooth curve. Do not go past the edge of your map. You have no data from which to make a pattern.
4. After your first line is complete, the next line will be easier because the lines tend to follow each other.

A contour map is called a **topographic map** because it shows the **topography** or shape of the land. A topo map, as it is often called, shows steep and gentle slopes by the spacing of the contours. It also shows valleys and rivers and most other features of the terrain.

Rivers on a contour map have a specific pattern that should be recognized. Whenever rivers are shown on a topographic map, the contour lines get bent by the shape of the valley.

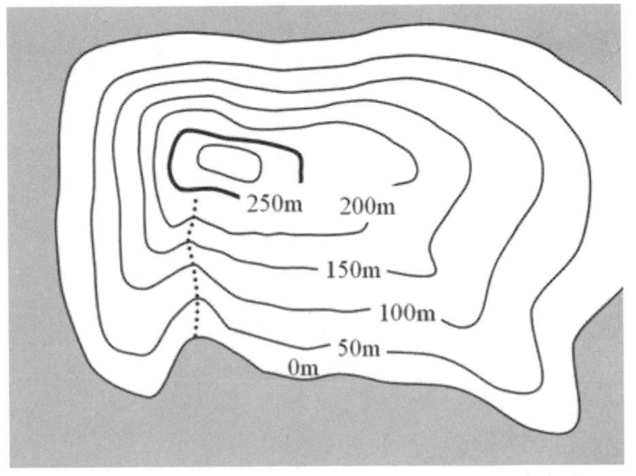

Figure 2.9 River

In Figure 2.9, the river is shown as the dotted line. The contour lines all bend towards the top of the mountain wherever they cross the stream. Because water always runs downhill, we can see that this stream is flowing southward. Another way to find the direction of flow of a river is simply to compare the values of the contours. Water always flows downward—in other words, from the higher elevation to the lower.

If you wanted to get a better picture of the steep and gentle areas on a map, you can construct a topographic "profile." Just as your profile shows what *you* look like from the side, a topographic profile shows how the *land* would look from the side. It's like slicing through the land and looking at it from the side as if it were a slice of cake. This is especially helpful if you are planning a hike through the area and want to plan for the terrain.

To Make a Topographic Profile

1. Draw a line across the map and label the ends A and B. This represents the "slice" through the land that you are about to make.
2. Place scrap paper along the line AB on your contour map. Be sure that this paper doesn't move until you complete the next two steps.

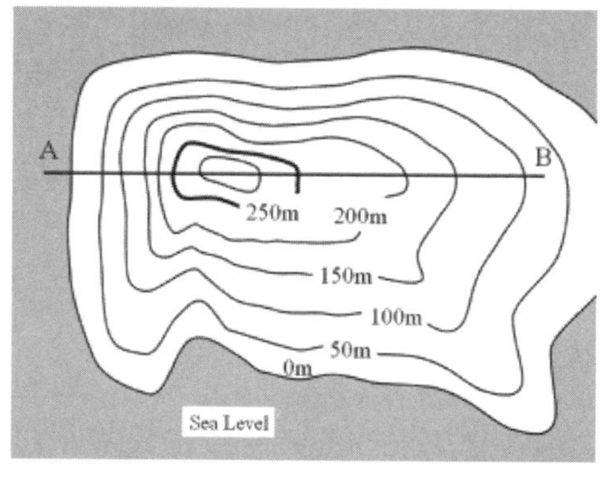

Figure 2.10

3. Put a "tick mark" wherever a contour touches the scrap paper. Be careful not to put tick marks for roads, rivers, or other kinds of lines.
4. Label the tick mark with the elevation of the contour line that the tick mark represents. It is okay to have several ticks in a row with the same value.
5. Take the scrap paper with the tick marks, and line it up at the bottom of a piece of graph paper.

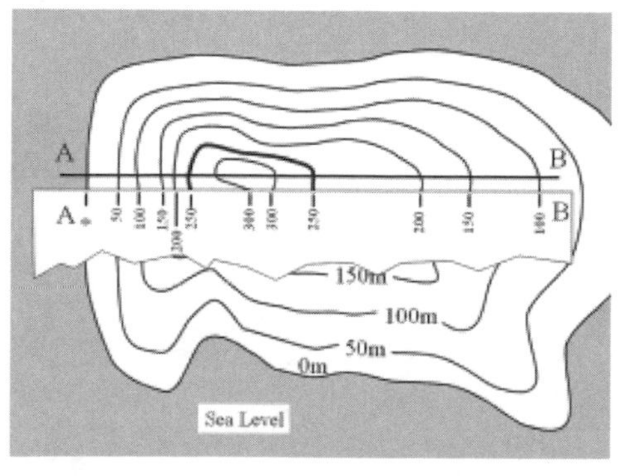

Figure 2.11

6. Put a dot directly above each tick mark at the elevation that it represents.
7. Connect the dots with a smooth curve in their correct order from A to B. However, whenever your smooth curve crosses a dot it must cross that elevation line. If you have two dots in a row of the same elevation, you will either have a hill between the dots or a dip.

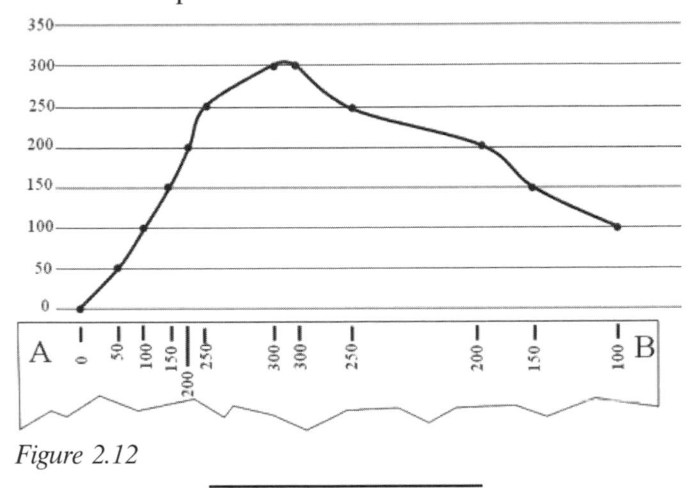

Figure 2.12

Maps

A map is a drawing of the land that gives the reader a perspective on distance and direction. However, there is one fundamental problem that all flat maps must contend with: To take a spherical object (the Earth) and show it on a flat piece of papcr, something has to get distorted—the distance, the direction, or both. The good news is that the smaller the area of the Earth that the map shows, the smaller the distortions. Think of it this way: One way to make a map would be to take the paper off of a globe (the most accurate map there is in terms of distance and direction) and spread it on a tabletop. If your map is going to be of the entire Northern Hemisphere, you will need to make several cuts to the paper to get it to lay flat—which will distort distance and/or direction. But, if you make a map of Washington, D.C., then you only need to cut out a small section of the globe and you will see that the little square from the globe lies almost flat without any distortion at all.

The way that maps were made long ago was to take a clear globe with a light bulb in the middle and use it to project an image of the land onto a piece of paper. These **map projections** have different styles depending on how the paper was held against the globe.

One of the most popular map projections is the **Mercator** projection. The Mercator projection preserves true direction: all lines of latitude are straight left and right while longitudes are straight up and down. On this map, the distance gets distorted more and more as the map gets closer to the poles, which are not shown on these maps. An example of the distortion is to notice that Greenland and South America appear almost the same size. In reality, Greenland is much smaller.

A **gnomonic projection** distorts both distance and direction. Near the center of the map, the image is close to accurate but gets more distorted at the edges. One advantage of the gnomonic map is that it shows the shortest distance between two locations as a straight line. Because the Earth's surface is curved, the shortest distance is actually a curve over the surface called a great circle.

A **great circle** is any complete circle around the Earth with the core at its center. If you were to slice the world in half at the Equator, which is a great circle, you'd slice through the core. If you sliced through the Arctic Circle (not a great circle), you'd take the top off of the Earth and not slice through the core. Each of the meridians of longitude, along with their counterparts on the opposite side of the world, together makes a great circle. Long distance airline flights follow great circle routes because they are the shortest routes across the curved surface of the Earth. For example, a flight from NY to Moscow goes almost over the North Pole. To find a great circle route on a globe simply take a string and place either end at the take-off and destination points. Tighten the string and it will mark the shortest route—a portion of a great circle.

Magnetic vs. Geographic Poles

Due to the Earth's rotation, the world has natural "starting points" for navigation: the North and South Poles. Even though these are real features of the Earth, there are no physical objects at these locations to aide anyone navigating. Luckily for us, there is a close approximation that can be used to navigate: the magnetic poles. You'll hear people say that "a magnetic compass always points north." That's not exactly true.

The Earth has a magnetic field that is almost lined up perfectly with the axis of rotation. As a result, magnetic compasses point almost to the North Pole, giving us a sense of the general direction. The **Magnetic North Pole**, the place where compasses point, is actually a spot on an island in extreme northern Canada. If you follow your compass, that's where it will take you.

Because these two "north poles" are not exactly together, there is some error in using a magnetic compass. This is corrected by setting the **magnetic declination** of the compass. This is a correction value given on all navigation maps. Once set with the magnetic declination, a compass will point to the true North Pole also known as the "**Geographic North Pole**." All places on Earth have a slightly different magnetic declination depending on how you line up with both poles. In an extreme case, if you stood between the two poles and faced the real North Pole, your magnetic compass would point to the magnetic North Pole behind you—a magnetic declination of 180 degrees!

Features of a Topographic Map

Contour Interval

The **contour interval** is the amount that the elevation changes from one contour line to the next. Sometimes it will be indicated with a simple "CI=."

Bold 5th Line

On contour maps, every fifth line is bold. This helps when counting lines that are either very close to each other or very far apart.

Magnetic Declination

The value of the **magnetic declination** represents how "wrong" a magnetic compass will be in that region of the world. A magnetic compass will point to the Magnetic North Pole and not the real North Pole. The magnetic declination will state what the correction will be so that the compass points at the Geographic North Pole. The magnetic declination can be given as a number of degrees or as a drawing of an angle labeled with North (sometimes just shown as a star) and Magnetic North (sometimes "MN").

Scale

The **scale** of a map tells how far it is from one location on a map to another. It is often shown several ways on each map.

A graphic representation is simply a ruler with the distances marked. They often are drawn in miles, kilometers, and feet on three separate scales on the same map.

Under the rulers of the graphic scales a ratio is usually printed such as 1:1,000. This means that one inch on the map is 1,000 inches in the real world. It also means that 1 cm = 1,000 cm; 1 foot = 1,000 feet; 1 finger-length = 1,000 finger-lengths.

Sometimes a really odd ratio—1:63,360—will be printed. This number seems random but it actually means that 1 inch = 63,360 inches, or 1 mile.

Lesson 2–5 Review

Exercise A: Drawing Isolines

a. Use a contour interval of 10 m.

b. Put a dot at all data points with the same value.

c. Assume that the values were measured at the center of each data point.

d. Connect the dots with a smooth curve.

42	45	50	51	49	48	48	47	46	44
44	47	51	55	51	49	48	48	48	45
46	49	55	57	56	50	49	49	50	45
47	50	56	62	60	54	50	53	54	50
48	55	60	65	63	57	54	58	57	53
50	57	64	67	67	60	58	62	60	55
52	58	67	73	73	65	60	63	60	55
53	60	69	78	80	70	65	68	60	55
54	60	70	79	85	72	72	74	64	54
55	61	69	79	90	75	80	75	66	55
55	60	67	77	80	75	80	75	65	55
53	60	67	73	75	72	75	70	65	55
52	57	65	68	68	68	69	69	61	54
50	56	62	64	63	61	60	64	60	54
48	51	58	57	59	50	48	54	56	53
45	47	52	53	51	49	46	49	51	51
43	45	48	46	49	46	44	45	49	48

Exercise B: Drawing Profiles

1. Complete the following tasks on the map shown in Figure 2.13. The map has a contour interval of 100 ft.
   a. Label every contour line.
   b. Draw the profile of the land along line AB.
   c. Draw the profile of the land along line CD.

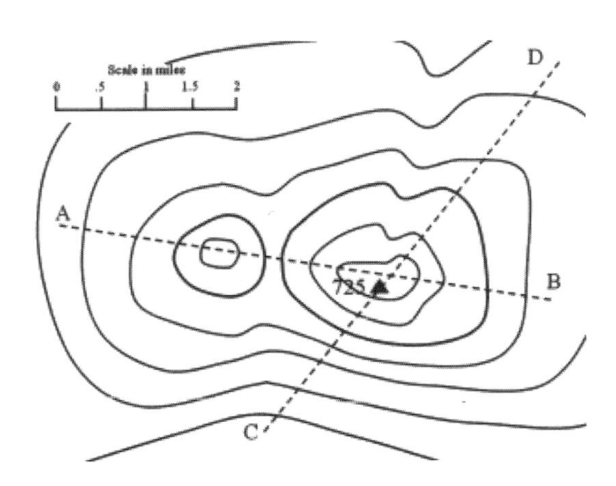

Figure 2.13

Multiple Choice

1. There is a stream in the northeast corner of the map above which is not drawn but you can see its fingerprint. Which way is the stream flowing?
   a) north b) east c) south d) west
2. Explain where the steepest part of the terrain is located on the map.
3. Calculate the gradient from C to the peak of the mountain.
4. Calculate the gradient from D to the peak of the mountain.
5. What does a directional compass point to?
   a) Magnetic North
   b) Geographic North
   c) the Prime Meridian
   d) the Equator

Chapter Exam

Exercise A

Record the coordinates of locations A through E on the following map (Figure 2.14).

A. ______________

B. ______________

C. ______________

D. ______________

E. ______________

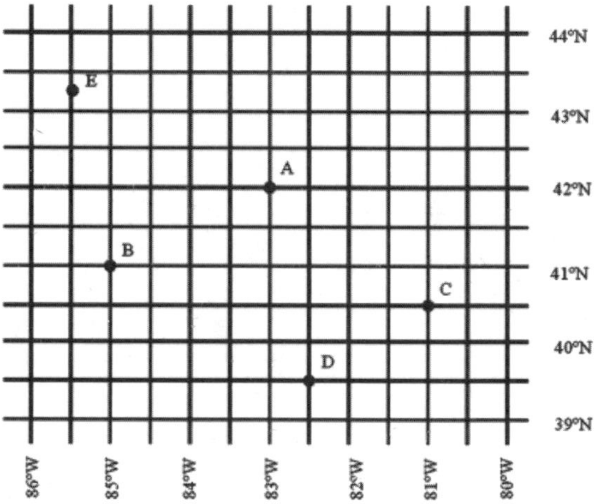

Figure 2.14

Exercise B

Construct contour lines with a contour interval of 10 m. Assume that the values were taken at the centers of the numbers.

13	16	19	21	22	22	23	23	23	23	20	17	15	12	9	5
16	19	21	24	26	26	25	24	25	23	21	17	16	13	10	8
19	21	25	27	28	29	30	29	28	24	21	18	16	14	13	11
20	25	28	30	33	36	35	34	29	25	22	18	16	15	14	13
21	25	30	34	38	42	41	40	30	26	22	20	18	17	15	13
20	23	28	30	34	40	42	40	33	28	26	23	21	20	20	19
19	20	24	26	29	30	38	37	34	31	30	29	28	27	25	23
18	19	20	22	25	28	31	34	36	36	35	34	32	30	26	24
14	15	18	20	24	26	30	37	41	41	41	40	37	31	27	24
11	13	15	18	20	25	30	36	42	44	44	42	37	32	26	24
9	10	12	17	20	25	29	35	41	43	43	41	37	31	26	21
6	8	10	15	20	24	28	33	38	40	40	39	34	29	25	20
6	8	10	14	19	22	26	29	30	30	30	29	28	26	22	19

Multiple Choice

1. The planet Neptune has a polar diameter of 48,682 km and an equatorial diameter of 49,528 km. Calculate Neptune's roundness.

2. One of Mars' two moons is called Phobos and is shaped like a potato. Calculate Phobos' roundness ratio if it is 9.2 km wide in one direction and 13.4 km in the other direction.

3. Which drawing shows the most accurate representation of Earth's true shape?

   a) 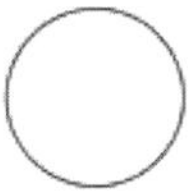 b) 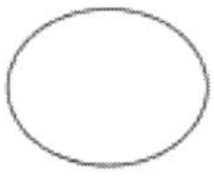 c) 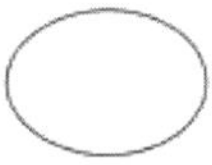 d)

4. A newly discovered planet is only partially studied. The distance between two locations on its surface is 83 km, and the change in latitude between the two locations is 20°. If both locations are directly north and south of each other, what is the circumference of the planet?

5. Calculate the circumference of this circl (Figure 2.19).

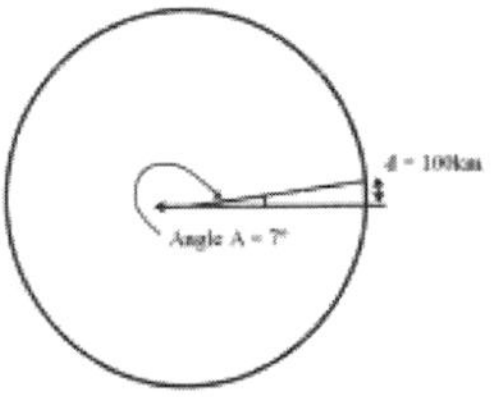

6. Your local time is 5 hours earlier than Greenwich, England's time. What is you latitude?

   a) 5°W c) 75°E
   b) 15°W d) 75°W

Figure 2.19

7. While on vacation, an observer travels to a location that is 12° north of the equator. Where will she see Polaris at night?
   a) directly above her
   b) right on the northern horizon
   c) 12° above the northern horizon
   d) 78° above the northern horizon

8. Your local time is 5 p.m. while it is 11 p.m. in Greenwich, England. Polaris is 36° above the horizon. What are your coordinates?

9. Using the scale shown here, how many kms wide is this book?

   0 km [________________] 1 km

10. Using the same scale, how tall is this book in kms?

Answer Key

Answers Explained Lesson 2–1

1. $\text{roundness ratio} = \dfrac{\text{polar diameter}}{\text{equatorial diameter}}$

2. $\text{roundness ratio} = \dfrac{\text{polar diameter}}{\text{equatorial diameter}} = \dfrac{133{,}708\ \cancel{\text{km}}}{142{,}984\ \cancel{\text{km}}} = \mathbf{0.935}$

   Note: The units cancel out so the answer has no kms after the number.

3. $\text{roundness ratio} = \dfrac{3{,}472\ \cancel{\text{km}}}{3{,}476\ \cancel{\text{km}}} = \mathbf{0.9988}$

4. $\text{roundness ratio} = \dfrac{6{,}750\ \cancel{\text{km}}}{6{,}794\ \cancel{\text{km}}} = \mathbf{0.9935}$

5. $\text{roundness ratio} = \dfrac{1{,}682\ \cancel{\text{pixels}}}{1{,}686\ \cancel{\text{pixels}}} = \mathbf{0.9976}$

Answers Explained Lesson 2–2

1. $\dfrac{A}{360°} = \dfrac{D}{C}$ or $C = \dfrac{360°}{A} \times D$

2. $C = \dfrac{360°}{A} \times D = \dfrac{360°}{15°} \times 30\ \text{m} = \mathbf{720\ m}$

3. $C = \dfrac{360°}{A} \times D = \dfrac{360°}{45°} \times 6\ \text{in} = \mathbf{48\ in}$

4. $C = \dfrac{360°}{110°} \times 20\ \text{m} = \mathbf{65.45\ m}$

5. $C = \dfrac{360°}{10°} \times 44\ \text{mm} = \mathbf{1{,}584\ mm}$

Answers Explained Lesson 2–3

1. From top to bottom, the major layers are: the atmosphere, hydrosphere, crust, mantle, outer core, and inner core.
2. The lithosphere is made of the crust and upper mantle.
3. **A** is incorrect because the mantle is under the crust.

   B is **correct** because the oceans sit on top of the crust and at the bottom of the atmosphere.

C is incorrect because the oceanic crust is part of the crust.

D is incorrect because the lithosphere includes the crust and part of the mantle below.

Answers Explained Lesson 2–4

1. "Internal angle" means the angle that would be made if you were at the center of the Earth and pointed one hand at the equator and the other at some other location on the surface. The furthest that you can point away from the equator is 90°, which is at one of the poles. If you point more than 90° then you will be making a smaller-than-90° angle on the other side of the Earth.

   A, **B**, and **D** are incorrect.

   C is **correct**.

2. **A** is **correct** because the North Pole is 90° north of the equator.

   B is incorrect because the equator is 0°.

   C is incorrect because this is a longitude—a measure of how far east or west you are from England.

   D is incorrect because latitude is degrees away from the "equator." If you pointed 180° away from the equator, you would be pointing back at the equator on the other side of Earth. The largest latitude is 90°.

3. **A** is incorrect because 90°is the spot directly above this observer. There is no star there.

   B is incorrect because 60° is the angle between Polaris and the spot directly above the observer.

   C is **correct** because the angle between the horizon and Polaris is 30°.

   D is incorrect because this would mean that Polaris would be right on the horizon.

4. **D** is **correct**. The furthest east or west that you can get from England is 180°. Once you go further than that, you start getting closer on the other side of the world.

   A, **B**, and **C** are incorrect.

5. **A** is **correct** because this is the definition of longitude.

   B is incorrect because lines of latitude are called parallels, not meridians. But it is true that latitude lines never touch.

   C is incorrect because the Prime Meridian is only half of a circle. It stops at the poles. The International Date Line (180°) is the continuation of the Prime Meridian.

   D is incorrect because this is the definition of longitude.

Answers Explained Lesson 2–5

Exercise A

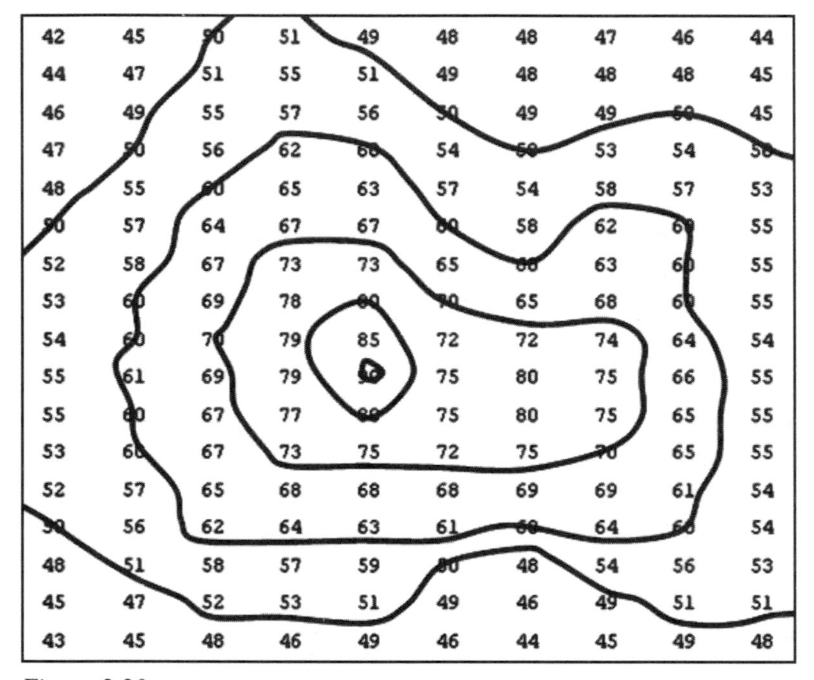

Figure 2.20

Exercise B

a.

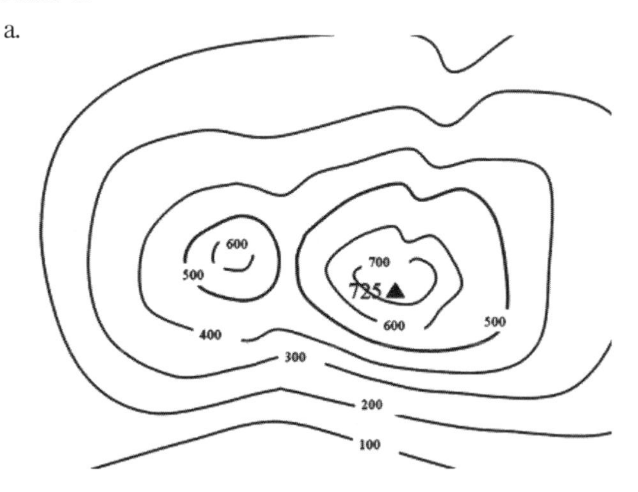

Figure 2.21 Labels

b.

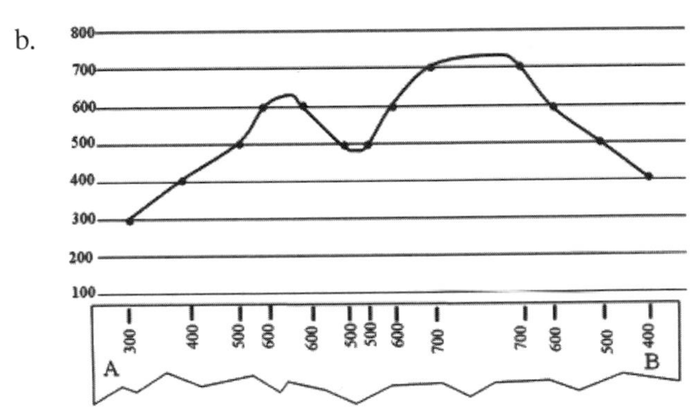

Figure 2.22 AB

c.

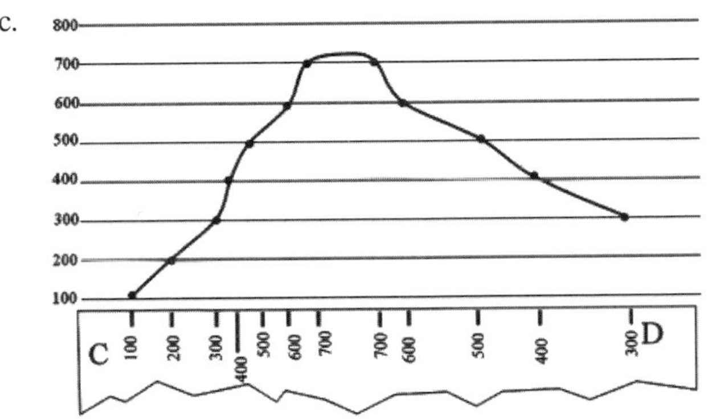

Figure 2.23 CD

Mulitple Choice

1. **A** is **correct** because the contour lines bend towards the peak of the mountain. The river is flowing from south to north. You can also compare the contour lines: the peak of the mountain is about 700 feet and the other end of the stream is about 100 feet high.

   B is incorrect.

   C is incorrect. Read the explanation to answer a. If you chose this answer because you've heard that all rivers flow south (at least in the Northern hemisphere), it is not true. Rivers flow downhill. North and south are not the same as up and down.

   D is incorrect.

2. The steepest part of the terrain is located on the southern side of the eastern mountain (between the 725 ft peak and "C"). It is the area where the contour lines are closest.

3. $$\text{gradient} = \frac{\text{change in value}}{\text{distance}} = \frac{725 \text{ ft} - 75 \text{ ft}}{2 \text{ miles}} = \mathbf{325\ ft/mi}$$

4. $\text{gradient} = \dfrac{\text{change in value}}{\text{distance}} = \dfrac{725 \text{ ft} - 150 \text{ ft}}{3.1 \text{ miles}} = \mathbf{185\ ft/mi}$

5. **A** is **correct** because a compass does not point to the true North Pole. It points to the strongest section of Earth's magnetic field—the Magnetic North Pole.

   B is incorrect because a compass does not point to the real North Pole of Earth.

   C and **D** are incorrect because a compass points to the Magnetic North Pole.

Answers Explained Chapter Exam

Exercise A

a) 42°N, 83°W

b) 41°N, 85°W

c) 40°30′N, 81°N (40.5°N is also acceptable because many GPS systems use decimals.)

d) 39°30′N, 82°30′W (39.5°N 82.5W)

e) 43°15′N, 85°30′W (43.25°N 85.5°W)

Exercise B

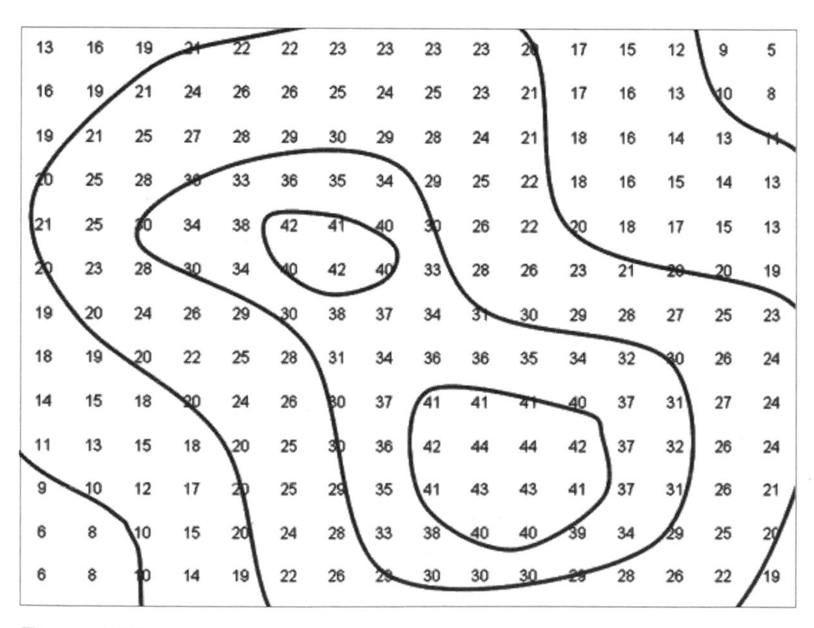

Figure 2.24

Multiple Choice

1. $\text{roundness ratio} = \dfrac{\text{polar diameter}}{\text{equatorial diameter}} = \dfrac{48{,}628\ \cancel{\text{km}}}{49{,}528\ \cancel{\text{km}}} = \mathbf{.982}$

2. $\text{roundness ratio} = \dfrac{\text{polar diameter}}{\text{equatorial diameter}} = \dfrac{9.2\ \cancel{\text{km}}}{13.4\ \cancel{\text{km}}} = \mathbf{.697}$

3. **A** is **correct** because Earth's roundness is so close to a perfect circle that we cannot see any difference. (This drawing could be a perfect circle or it could be a very slightly squished one—we can't see the difference.)

   B is incorrect. Even though this drawing shows a very slight oblateness, it is still way to exaggerated to accurately represent the Earth. (This oval has a roundness of .80 while Earth's is .997.)

   C and **D** are incorrect.

4. $C = \dfrac{360°}{A} \times D = \dfrac{360°}{20°} \times 83\ \text{km} = \mathbf{1{,}494\ km}$

5. $C = \dfrac{360°}{7°} \times 100\ \text{km} = \mathbf{5{,}142.9\ km}$

6. **A** is incorrect because each hour of time difference represents 15°.

   B is incorrect because each hour time difference represents 15°. This would be the latitude if the time difference was only one hour.

   C is incorrect because you have the correct degrees but the incorrect direction. See the explanation below.

   D is **correct** because each hour of time difference represents 15° of latitude. 5 × 15° = 75°. Traveling west, the time gets earlier (it is earlier in California than it is in New York).

7. **A** is incorrect because the only place on Earth to see Polaris directly above you is at the North Pole.

   B is incorrect because in order to see Polaris on the horizon, you must be at the equator—latitude 0°.

   C is **correct** because your latitude is the same as the angle that Polaris will be above the horizon. It is always in the northern direction.

   D is incorrect because 78° is how far away Polaris will be from the spot directly above her in the sky.

8. 36°N, 90°W

   The angle to Polaris is your latitude.

   Your longitude is the difference between your local time and Greenwich time multiplied by 15°.

11 p.m. – 5 p.m. = 4 hours

$4 \times 15° = 60°$

Time gets earlier as you go west (California is earlier than New York) so your longitude is 60°W.

9. 4.8 km. Take a scrap piece of paper, mark the length of the scale on the scrap and see how many scaled kilometers will be needed to span the book.
10. 7.2 km. See the previous explanation on how to measure with a scale.

Minerals and Rocks

Lesson 3–1: Minerals

A **mineral** is a naturally occurring, inorganic solid with a definite chemical composition and crystalline structure. Of the list of requirements just listed, probably the two most important are the chemical composition and crystalline structure. These two criteria are the final judges of a mineral's identity. The chemical composition refers to the mineral's formula. For example, quartz's formula is SiO_2, meaning that one unit of quartz is made of one silicon atom joined with two oxygen atoms.

Mineral Identification Tests

Color

Color is the first test done to identify a mineral, but it can also be the least reliable. Even before you pick up a sample, you are already looking at its color and your brain starts working on identifying the mineral. The color test is not always accurate for two main reasons: Many minerals have the same color and many minerals can be found in several colors. The two most common colors in minerals are white and clear. There are dozens of minerals that share a white or clear color.

To make things more difficult, there are many minerals that can appear in more than one color. The variety of colors is caused by impurities in the mineral. An impurity is a trace chemical that creates a color but there is not enough of the chemical to change the chemical composition of the mineral. An example would be using food coloring to change a vanilla cake to a green cake for St. Patrick's Day. The cake still tastes the same but is no longer the ordinary color.

Streak

The **streak** of a mineral is the color of the powdered mineral. The streak is obtained by rubbing the unknown mineral on a streak plate, which is an unglazed porcelain tile. The color of the streak can be different than the mineral's color because the light passes through the powder rather than bouncing off of the mineral's surface. With many minerals, the streak test can be useful for identification. Iron pyrite, otherwise known as Fool's Gold, is easily identified by its streak. The mineral appears gold but the streak is a definite black or dark green. By contrast, real gold has a gold streak.

The streak test is useless for clear or white minerals. All of these minerals will invariably have a white powder streak; therefore the results are already known. Minerals with a hardness greater than the streak plate (around 7 on Mohs scale) also cannot be tested with the streak test. Many minerals are not identified by their streak, but it is one piece of evidence used in identification.

Hardness

Hardness is a mineral's resistance to scratching. It is not to be confused with brittleness or density. For example, diamond, the hardest mineral known, can shatter if struck with a hammer. Similarly, you would not want a pillow made of one of the softest minerals, talc. Although talc is very soft and scratches easily, it is not fluffy at all.

Mohs Scale of Hardness

The system used today to rank a mineral's hardness was developed by Friedrich Mohs. He created a scale from 1 to 10 using common or easily obtainable minerals to aid geologists in the field. In addition, there are some simple tools that a field geologist may have in her backpack that are also ranked on the scale.

The minerals from softest to hardest are:

Mohs Scale of Hardness

1. Talc.
2. Gypsum.
3. Calcite.
4. Fluorite.
5. Apatite.
6. Feldspar.
7. Quartz.
8. Topaz.
9. Corundum.
10. Diamond.

Key Points of Doing a Hardness Test

You can determine the exact hardness range of a mineral if you have a few minerals of known hardness. Otherwise, you can place minerals in order of relative hardness.

- Choose one mineral to be the scratcher and one to be the scratchee.
- Pick a smooth, flat surface on the scratchee to get scratched.
- After doing the test, wipe the powder away to confirm that the scratchee really was scratched.
- If the scratchee did not get scratched, switch the two rocks and repeat.

  It is important to wipe the powdery scratch away and look for an actual scratch in the surface. What appears to be a scratch may actually be the crumbled powder remains of the scratcher.

Luster

Luster is the way that a mineral shines or doesn't shine. The two main divisions in luster type are metallic and non-metallic. The **metallic** luster shines the way a polished piece of metal does. Some familiar examples of a metallic luster are gold and silver jewelry, a clean iron nail, an aluminum can, and a chrome water faucet. To identify minerals using the luster test, it is best to view a freshly broken sample. Weathering or deposits on the surface may affect the luster. However, samples in your class are not exposed to weathering and will usually display the true luster.

Non-metallic minerals, simply put, do not shine like a metal. Because there are many ways that something can look different than a metal, there are many kinds of non-metallic lusters. The best ways to know these lusters are to actually hold and see a sample displaying that luster.

Some common non-metallic lusters:

- Glassy, also known as **Vitreous.**
  - Probably the most common non-metallic luster.
  - Looks like a piece of broken glass.
- Waxy, also know as **Resinous.**
  - The sample appears to be made of plastic.
  - Sometimes the edges will look as if clumps of wax are stuck to the corners.

- Pearly.
  - A pearly luster looks like the inside of clam and oyster shells.
  - It will look like the surface is oily.
  - It may have a rainbow effect to the shine.
- Dull or earthy.
  - The sample has no shine to it at all.
  - Dark samples will look like dirt.
  - Light samples will look like chalk.

Cleavage

Cleavage is the way that a mineral will split or break along flat planes. A mineral that cleaves will often have smooth, flat, reflective surfaces. The shape of the cleaved mineral comes from the way that the crystal's atoms and molecules line up to form zones of weakness.

Cleavage planes are the flat surfaces produced by splitting along the zones of weakness. Minerals can have one, two, or several cleavage planes. It is similar to chopping potatoes for cooking. Each plane is a different direction you have to turn the potato to cut. For example, to make slices, you don't turn the potato and make one series of parallel slices. To make diced potatoes, you cut a series of slices, turn the potato and make a second series of slices and then turn the potato again to complete the three directions of the cubes.

Perfect vs. Imperfect cleavage. Any type of cleavage (basal, cubic, and so on) can be perfect or imperfect. Perfect cleavage produces smooth, flat, reflective surfaces. Imperfect will make what, at first, appears to be a rough surface. However, if you turn the sample in the light, one entire side will flash at once even though it is not flat. This is caused by a stair-stepping effect: The sample breaks along small parallel planes, which all reflect light at the same angle. In addition, minerals that exhibit cleavage can sometimes show fracture patterns.

Types of Cleavage

- **Basal**—to split into flat sheets. The best example of basal cleavage is mica. Mica can be split with a fingernail and will peel into extremely thin transparent sheets and looks like rigid plastic wrap. Sometimes basal cleavage is referred to as splitting along one plane or in one direction.

- **Cubic**—to break apart into cubes. Technically speaking, cubic cleavage results from a "mineral having three different cleavage planes that are mutually perpendicular." Put simply, all the sides of a cubic mineral meet at right angles. It produces cubes and square patterns. Two examples of cubic cleavage are halite and galena.
- **Rhombohedral**—similar to cubic except the cubes are pushed over a little. The three cleavage planes do not meet at right angles. Instead of cubes, this will produce a rhombus-shaped cube: a rhombohedron. Calcite demonstrates excellent rhombohedral cleavage.

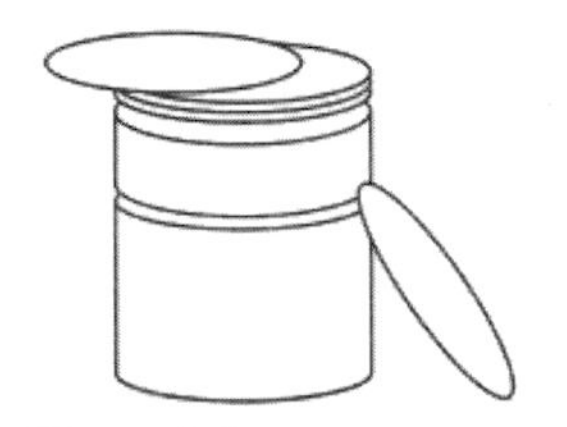

Figure 3.1
Basal Cleavage

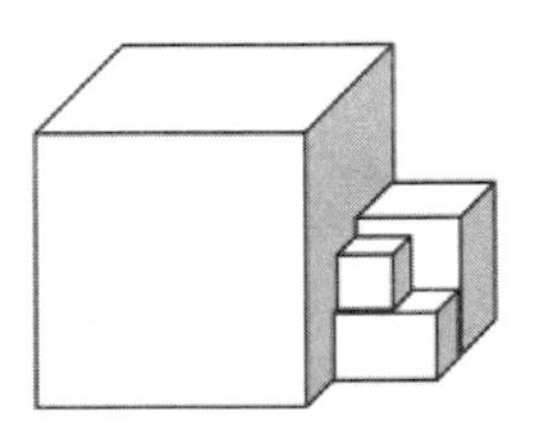

Figure 3.2
Cubic Cleavage

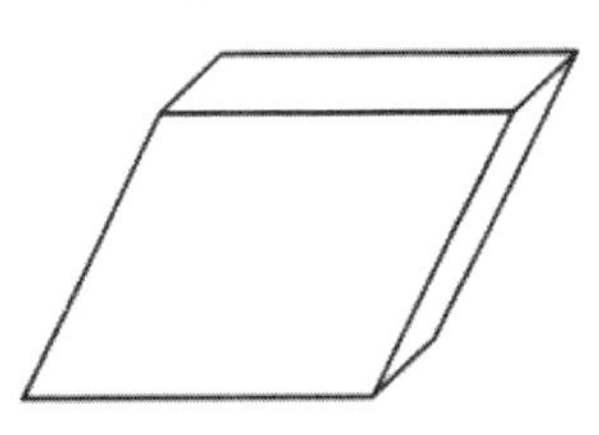

Figure 3.3
Rhombohedral cleavage

Fracture

Fracture is the way a mineral without cleavage breaks. Although fracture may be a lack of cleavage, it may also give some useful information for mineral identification. Some minerals have distinctive fracture patterns.

Types of Fracture

- **Splintery fracture**: Little needles break off the mineral. Typical of hornblende.
- **Conchoidal fracture**: The mineral breaks into smooth, "scooped-out" bowl depressions. Named after the *conch* sea snail's shell. This type of fracture is seen in some samples of quartz and is probably seen best in obsidian (igneous rock).

Miscellaneous Tests

There are some tests that will be helpful if you suspect that you have a *specific* mineral. These tests will help confirm your suspicions.

The Acid Test: Used for Calcite and Powdered Dolomite

A drop of diluted hydrochloric acid (HCl) will fizz when placed on the mineral. Sometimes the fizzing will be too slow to see so it may be helpful to hold up to your ear to hear the fizzing. (Be careful not to touch the acid to your ear or hair.)

Smell: Sphalerite

Sphalerite will give off a rotten egg smell when you do a streak test with it.

Magnetism: Magnetite (lodestone)

As the name implies, magnetite is magnetic. You will often see tiny bits of metal stuck to its surface. Magnetite will pick up a paper clip. If you are testing a weak sample of magnetite a small staple may be needed with which to do the test.

Taste: Halite

Halite is otherwise known as rock salt. It will have the distinctive taste of table salt. However, be sure to check with your instructor if it is safe to do the taste test. The samples have been handled by other students and may have residual acid from the acid test on them.

All Physical Properties of Minerals Come from the "Internal Arrangement of Atoms."

An excellent example of how the arrangement of atoms affects the properties of a mineral is the contrast between graphite and diamond—both of which are pure carbon. Graphite is one of the softest minerals known; diamond is the hardest. Graphite is dark and opaque, diamond is clear and colorless. The only difference between these two minerals is the pattern of the atoms.

A Spotlight on Quartz

Quartz is one of the most common and important minerals to us. It has some remarkable properties that deserve some special attention.

Quartz is one of the most abundant minerals on Earth. If you pick up a pebble on the ground, chances are it is a quartz pebble. It is so common that you have some at home and you are probably wearing some right now!

Quartz's chemical formula is SiO_2. Because it is one of the most abundant minerals, silicon and oxygen are the two most abundant elements in the crust—oxygen being the number one element. It is strange to think that the solid rock beneath us is made mostly from oxygen but we must keep in mind that when elements combine with other elements, the original properties usually change.

Quartz's structural unit is called the **tetrahedron** which is a three-sided pyramid (plus the base makes four sides) with the oxygen occupying the corners and the silicone atom in the center of the pyramid. Initially, it appears as if it should be SiO_4, but the oxygens at the corners are also being shared by neighboring tetrahedrons (reducing the amount of oxygen from four down to two). Overall, in a sample of quartz, there are twice as many oxygen atoms as silicons.

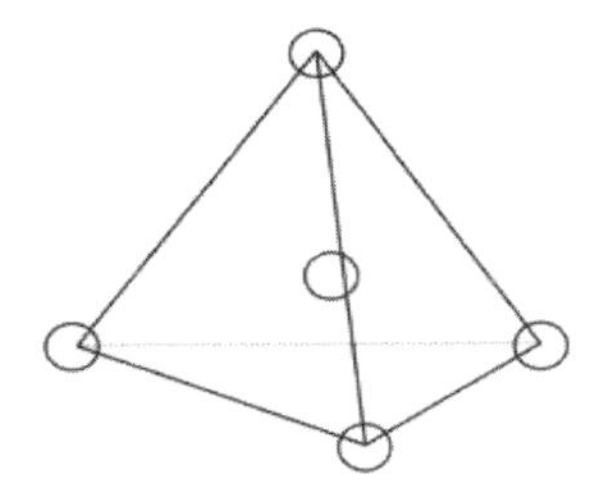

Figure. 3.4 *Tetrahedron*

Quartz is very hard (7 on Mohs scale) and chemically resistant. For this reason it is one of the last minerals to be worn away when a mountain crumbles. As a result, the most common sediment is sand made from bits of quartz. When sand is melted and resolidified, it forms glass. Glass is an ideal material because it has quartz's properties of hardness and chemical resistance—which explains why it is used in lab equipment to hold chemicals such as acid. Quartz's hardness is also the reason why it is used as an abrasive in *sand*paper.

Another interesting use of quartz is in watches. Inside every "Quartz" watch is a tiny crystal of quartz. The crystal vibrates with such precision that it is used to tell time. The vibrations act like a pendulum, which the watch counts to keep track of time.

Lesson 3–1 Review

1. A mineral's resistance to scratching comes from
   a) its luster.
   b) its internal arrangement of atoms.
   c) the shape of the mineral sample.
   d) fracture pattern.

2. White is not a reliable color to use for mineral identification because
   a) white is a rare color.
   b) white is a common color.
   c) most people cannot tell the different shades of white apart.
   d) all white minerals are quartz.

3. Splitting into thin flat sheets is known as
   a) cubic cleavage. c) basal cleavage.
   b) imperfect cleavage. d) feldspar.
4. A mineral's hardness can be tested by
   a) scratching it against another mineral.
   b) crushing it in a vise.
   c) hitting it with a hammer.
   d) splitting it into thin sheets.
5. The acid test is useful for identifying
   a) any non-metallic mineral. c) gold.
   b) calcite. d) hardness.
6. Another term for a glassy luster is
   a) metallic. b) resinous. c) waxy. d) vitreous.

Lesson 3–2: Rock Formation

Rocks

A rock is a material that is made from a mixture of minerals. Because it is a mixture, the amounts of each of the raw ingredients, the minerals, can vary greatly.

A **monomineralic** rock is a rock made of only one mineral. In this case, the rock is both a rock and a mineral. This duality confuses some people. This is comparable to a person being both a student and an athlete at same time. It all depends on the context of the conversation. If you are discussing rocks, then a pure sample of calcite would be considered a rock. If you are discussing minerals, then the calcite would play its role of a mineral. A good example of a monomineralic rock would be the common quartz pebble.

If a rock is made of more than one mineral, it is called a **polymineralic** rock. Polymineralic rocks tend to be more colorful due to the presence of the differently-colored minerals. Granite, with its "chocolate chip ice cream" speckles, is a fine example of a polymineralic rock.

Rock Cycle

Rocks tell a story. We just need to know the language of rocks in order to understand the story.

Here is a simplified story of the rocks on Earth: Earth formed from a ball of melted minerals during the birth of the solar system. As these minerals cooled, they solidified at the surface forming a shell of solid igneous rock. While the Earth's crust was forming, volcanic activity spewed out gases and steam to create the first atmosphere and oceans and therefore the first weather. Exposure to the weather generated on the surface of the planet wore down the igneous rock into sediments. The sediments collected into the low areas and began to accumulate as the mountains wore down. In some places the piles of sediment stacked up for hundreds or thousands of feet. The intense weight of the sediments compressed the bottom of the pile. While the compaction was going on, minerals dissolved in the groundwater between the sediments began to coat the sediment grains as mineral cement. The combination of **compaction** and **cementation** caused the sediments to "glue" together into solid sedimentary rock in a process called **lithification**. The **sedimentary** rock formed at the bottom of the stack of accumulating sediments is driven deeper into the earth under increasing pressure and temperature. The temperature may get so hot that the minerals begin to approach their melting points. The atoms in the rock rearrange and the structures distort forming new minerals. The result is **metamorphic** rock. Even though a new rock has formed, the process of heating and pressurization doesn't stop. Eventually, the minerals reach their melting points and the rocks form liquid magma. The return of the rocks to melted minerals completes the cycle and the rocks go on to become igneous, then sedimentary and metamorphic rocks again.

Although this is a typical story of how rocks can move through the rock cycle, it is not the only possibility. For example, a metamorphic

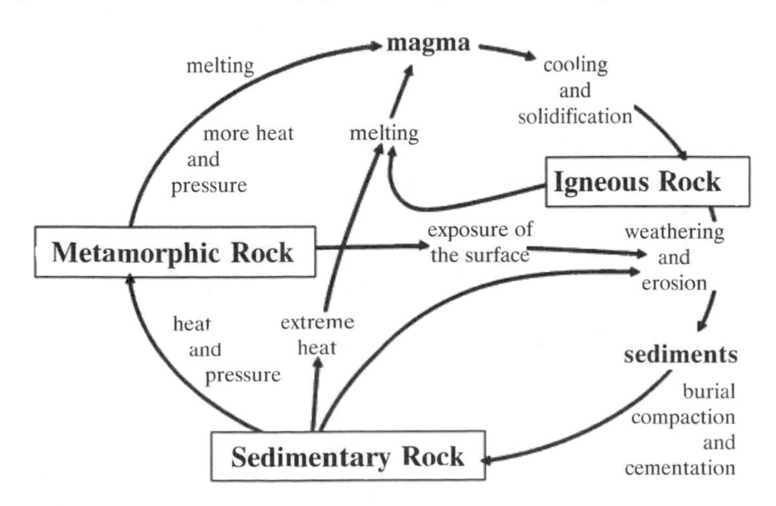

Figure 3.5 *Rock Cycle*

rock may become exposed to weathering at the surface. In this case, the metamorphic rock will become sediments and then sedimentary rock. The world is big enough and old enough to allow just about any variation of travel around the rock cycle.

Mineral and Rock Resources

In order for humans to use mineral, the minerals must first be mined. The minerals can be found in their pure form or in a form in which industrial processes are needed to extract the minerals. If a process is necessary to extract a specific mineral, the source rock is called an **ore**.

Coal, for example, is a resource that is found in its pure form. All that needs to be done in this case is to dig the coal out of the earth. Mining can be done as shaft mining and open pit mining. **Shaft mining** is digging a hole into the earth and tunneling to extract the materials. The ore is extracted using rail cars or elevators. Open pit mining exposes the deposit by removing all of the overlying material. **Open pit mining** tends to be controversial because it completely destroys the environment above the deposit. Responsible open pit miners go to extreme lengths to restore the environment when they are finished in the area.

When ore is extracted, the rock material must be processed to extract the target mineral. Aluminum, for example is not found in its pure form. It must be extracted from an aluminum-rich ore such as bauxite. After the mineral is refined from the ore, the waste material is either used by another industry or backfilled into the area where the ore was originally taken.

Mineral resources are **nonrenewable**, meaning that once they are used, they are gone forever or at least very long time. Prime examples of this are coal, oil, and other fossil fuels. It takes millions of years for these materials to be made by nature, and once we burn up the last of the resources they are gone. With modern concerns and technology, these resources are being used more efficiently. Cars get more miles per gallon and home heating is more efficient. In addition, recycling saves natural resources from being mined.

Lesson 3–2 Review

Use the diagram of the Rock Cycle on page 73 to answer the questions in this section.

1. As magma cools, it forms ________________ rock by the process of ________________.

2. Igneous rocks can form ______________ , ______________, and ______________ rocks.
3. Sediments form __________ __________ by the processes of ____________ and ________________ .
4. Sediments form from the process of ________________ .
5. Sedimentary rocks can form ______________ , ______________, and ______________ rocks.
6. Which process changes igneous rock into metamorphic rock?
7. Which processes changes sedimentary rock into igneous rock?
8. Which processes changes metamorphic rock into sedimentary rock?
9. Metamorphism involves the addition of ______________ and ______________ to preexisting rocks.
10. Compaction and cementation of sediments forms ______________ rocks.
11. Subjecting sedimentary rocks to extreme heat and pressure forms ______________ rocks.
12. Solidification of molten minerals forms ______________ rocks.

Lesson 3–3: Igneous Rocks

Igneous rocks are formed from the solidification of melted minerals. The name *igneous* comes from Latin and means "fire" as in "ignite." Deep inside the Earth, **magma** is made of a soup of minerals all mixed together. When the magma cools, the minerals will solidify into crystals. Sometimes the process of solidification is called "freezing" even though it is not happening at what we would call cold temperatures. As the crystals grow and get larger, they may bump into each other and become intergrown. This will appear like a puzzle made of geometric shapes. In contrast, a sedimentary rock will be made of cemented sediments that look like small particles packed together.

The Environment of Formation of Igneous Rocks

When an igneous rock has cooled and solidified deep within Earth, it is called an **intrusive** igneous rock or a **plutonic** rock. "Intrusive" refers to the magma that *intruded* in on the original rock as it melted its way towards the surface. A *pluton* is an underground formation of magma. Any rock formed within a pluton is a plutonic rock. Plutonic comes from Pluto, the Greek god of the underworld.

Igneous rocks formed at the surface of Earth will typically come from a volcano. These **volcanic** rocks are formed in an **extrusive** environment. When magma reaches the surface of Earth it becomes **lava**. The only difference between lava and magma is lava has reached the surface while magma remains underground.

The Texture of Igneous Rocks

While the magma cools, the crystals continue to grow as long as there is melted material available to solidify. If an igneous rock is made of large crystals (a coarse texture), it tells us that it took a long time to cool. Small crystals (fine texture) were produced in a short cooling time. If the cooling was so rapid that crystals do not have any time at all to form, the rock will have a glassy texture otherwise known as an "amorphous solid." Sometimes during an eruption, escaping gases will get trapped within the lava producing bubbles. The bubbles, called **vesicles**, give the rocks a vesicular texture.

The difference between texture and how a rock feels: The texture of igneous rock is strictly referring to the size of the crystals. A coarse texture is large crystals. A rock can have crystals so large that it will feel smooth if you touch the flat crystal faces. A fine-textured rock with small crystals can feel very rough if it has bubbles in it.

Mineral Composition

A good description of magma is that it is a "soup" of minerals. Each mineral can be present in varying amounts depending on the original rock that was melted. However, groups of minerals are often found together.

Felsic rocks contain a high percentage of *fel*dspar, hence *fel*sic. The felsic minerals contain aluminum and silica, which gives rocks formed from this group of minerals a light color and low density. One exception to the light color of felsic rocks is obsidian, which usually appears black. This is caused by small amounts of magnetite mixed throughout the lava. However, thin edges of obsidian will appear light in color.

The **mafic** minerals, which derive their name from *ma*gnesium and iron (*F*e), are usually dark colored and dense.

Igneous rocks can also be grouped into families. A family of minerals is a group of rocks with the same minerals but formed under different conditions. It's like using the same ingredients of flour, eggs, and sugar and creating many different recipes depending on how you cook them.

The **granite** family is the most felsic group and is made mainly from the minerals orthoclase feldspar and quartz. Granitic rocks are usually light in color and have a low density. The difference between each of the granitic rocks is how they are each made and not their composition. If the rock cools slowly, it will have large crystals and become granite. If it cools more quickly, it will have small crystals and form rhyolite. When granitic lava reaches the surface and cools very rapidly, it will be glassy obsidian. Bubbles trapped in granitic lava forms the vesicular rock pumice.

The **diorite** family of rocks consists mainly of plagioclase feldspar and the mafic minerals: hornblende (sometimes called amphibole) and biotite mica. A coarse texture will form diorite; the fine textured form is andesite. In the glassy samples of the diorite family, you will also find obsidian or pumice, depending on if there are bubbles present.

The **gabbro** family, which is highly mafic, is a dark dense group of rocks. This family contains high quantities of plagioclase feldspar and augite. The rocks may also contain olivine. The forms of gabbro from coarse to fine and glassy, respectively, are gabbro, basalt, and basalt glass—scoria, if there are vesicles.

Igneous Rock Chart

	Felsic ⟷ Mafic (low density) (high density)		
	Granite Family	Diorite Family	Gabbro Family
10mm Intrusive (Coarse)	Granite	Diorite	Gabbro
1mm	Rhyolite	Andesite	Basalt or Scoria
glassy Extrusive (Glassy)	Obsidian or Pumice		Basalt Glass

Figure 3.6

Now, keep in mind that magma and lava are mixtures, so the amount of each mineral can vary. As a result, the granite, diorite, and mafic families blend into each other at their edges. Likewise, there can be a variation in cooling rates resulting in overlapping textures.

Igneous Mineral Composition Chart

Figure 3.7 shows the "recipes" for many common types of igneous rocks. At first glance, it is a very complicated chart, but it is easily tamed. Listed at the top are many different igneous rocks separated by the size of the crystals (up-and-down) and the mineral composition (left-to-right). For example, granite and rhyolite have the same minerals in the same proportions—their only difference is the size of the crystals within each rock.

Composition Chart

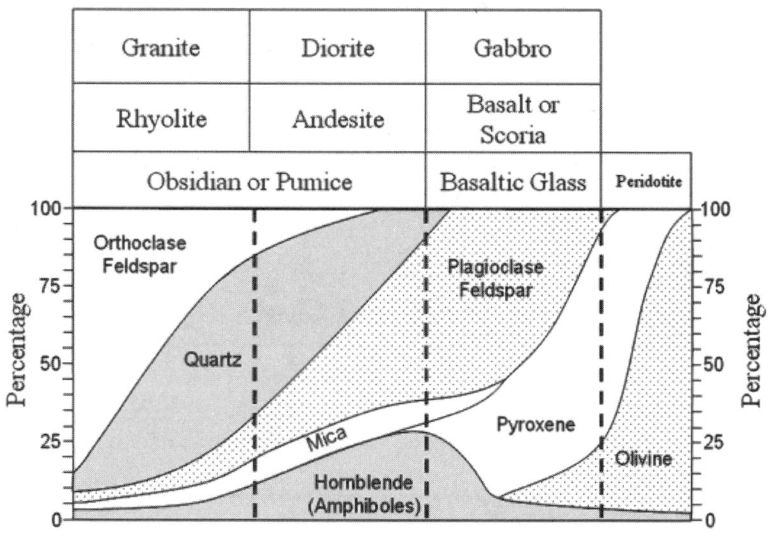

Figure 3.7

To Find the Minerals in a Given Rock

Simply take your finger, point it at the name of the rock and drag your finger straight down into the bottom half of the chart. Whatever "territories" your finger drags through are minerals present in the rock. For example, place your finger in the center of the word *rhyolite* and drag it straight down. Your finger will pass through rhyolite's minerals: orthoclase feldspar, quartz, plagioclase feldspar, mica, and hornblende. The sizes of the territories tell the relative percentages of each mineral in the rock.

There is no *exact* mixture for each rock. If you start your finger at the left side of the word *rhyolite* or the right side of the word, you will have slightly different amounts of each mineral. How far you go left or right depends on how felsic or mafic, light or dark, or dense the rock sample is. (See the Igneous Chart on page 77.) If you have no specific information as to the mineral's density, color, or mafic/felsic composition, start your finger at the center of the rock name and drag straight down.

To Find the Exact Percentages of Minerals in a Rock

If you want to know the percentages of minerals in the lightest sample of granite you would start your finger at the extreme left side of granite's box and drag your finger straight down. This one is easy because it goes right along the scale on the left of the chart. The amount of time that your finger spends in each territory tells you the percentage of each of the minerals. For example, the lightest granite contains about 85% orthoclase feldspar (from the 100% mark down to the 15% mark), 6% quartz (from 15% down to 9%), 3% plagioclase feldspar, 2% mica, and 4% hornblende. If everything is done correctly, you should end up with a total of percents close to 100%.

What if the sample is in the middle? Then you will use a scrap piece of paper and put tick marks on it. Try andesite:

- Start at the "e" in the middle of andesite.
- Line your paper straight up-and-down.
- Put a tick mark at the border of each territory your paper passes through (including the topmost and bottommost).
- Label each territory with the name of the mineral.
- Drag your paper over to the scale on the side of the chart and measure the percentages.

A middle sample of andesite will contain approximately:

- 4% orthoclase feldspar
- 35% quartz
- 30% plagioclase feldspar
- 9% mica
- 22% hornblende

Just to make things a little more interesting, it is actually possible to mix textures within one rock. This is caused by the minerals freezing or solidifying at different temperatures. As the magma cools, some minerals will begin to crystallize while all others in the mix will remain liquid. In the end, the first minerals to begin solidifying will have the longest time to grow. The result is a mixed texture known as a **porphyry**. The porphyritic texture can be described as large grains within a fine-grained matrix.

Lesson 3–3 Review

Exercise A

Use the Igneous Rock Chart on page 77 to fill in the blank spaces in the table.

Rock	Formed From	Cooling Rate	Crystal Size	Texture
Rhyolite	Lava		Small	
Gabbro		Slow		Coarse
Basalt		Fast		Fine
Pumice	Lava		No crystals	
Obsidian		Very fast		Glassy
Granite	Magma		Large	

Exercise B

Estimate the percentage of each mineral in gabbro (look under the first "b" in gabbro).

Olivine __________%

Pyroxene __________%

Plagioclase feldspar __________%

Biotite mica __________%

Hornblende (Amphiboles) __________%

Multiple Choice

1. Is granite intrusive or extrusive?
2. Plutonic is also referred to as
   a) intrusive. b) extrusive.
3. What is the difference between an igneous rock with a glassy texture and a fine texture? What causes this?
4. List the minerals present in andesite.
5. Compare rhyolite to basalt (answer in one word only).
   a) Which has a lower density?
   b) Which has a lighter color?
   c) Which has a more mafic composition?
6. Name a felsic, intrusive rock.
7. Felsic igneous rocks generally have a ____________ color.
8. One way to determine if an igneous rock has had an intrusive or extrusive origin is:
   a) Intrusive igneous rocks are always dark colored.
   b) Intrusive igneous rocks are more coarse-grained than extrusive rocks.
   c) Intrusive igneous rocks are more fine-grained than extrusive rocks.
   d) Intrusive igneous rocks are always less dense than extrusive rocks.
9. According to the Igneous Rock Chart (pg. 77), what is the texture of granite?
   a) glassy b) fine c) coarse d) vesicular
10. What is the grain size of granite?
    a) glassy c) between 1 mm and 10 mm
    b) less than 1 mm d) larger than 10 mm

Lesson 3–4: Sedimentary Rocks

In the truest sense, **sedimentary** rocks are made from sediments, or broken pieces of rock. However, this definition describes only one type of sedimentary rock. All sedimentary rocks have been deposited in some way—usually in a water environment (as opposed to igneous rocks, which are formed by solidification from melted material). Sedimentary rocks are sub-divided into three groups based on how they were deposited: clastic, chemical, and organic.

The clastic rocks are formed in the manner described in the rock cycle in lesson 3–2. **Clastic** rocks are the lithified (solidified into rock) accumulation of sediments. The sediments are the weathered remains of some preexisting rock (it can be igneous, sedimentary, or metamorphic). The sediments pile up and get pressurized by the weight of the sediment pile. It becomes a solid rock in a process called **lithification** when the pressurized sediments get cemented together by minerals dissolved in groundwater. In order for the sediment to become lithified into solid rock, both compaction and cementation are needed. Think of trying to throw a ball of sand at the beach. Pick up some dry sand and it falls through your fingers. Pick up some wet sand and it oozes between your fingers. But if you take wet sand and squeeze it, the ball will remain intact long enough to throw it at some unsuspecting friend. In this case, the squeezing is the compaction and the water is acting as a weak glue.

Clastic sedimentary rocks are identified only by their range of sediment sizes. Shale is comprised of clay particles, which are sediments that are smaller than .0004 cm—essentially rock bits the size of dust or flour. Siltstone and sandstone, obviously, are made of silt and sand particles. If the rock is made of larger particles, it will be a **conglomerate** or **breccia**. The only difference between the two is that conglomerate is made from rounded sediments while the breccia has more angular, less-abraded sediments.

Sediment Size	Sediment Name	Sedimentary Rock
Less than .0004 cm	Clay	Shale
.0004 cm–.006 cm	Silt	Siltstone
.006 cm–.2 cm	Sand	Sandstone
Mixed	Mixed	Conglomerate (rounded sediments)
Mixed	Mixed	Breccia (angular sediments)

Because clastic rocks are deposited as sediments settle down, they will often be layered. In addition, any dead plant or animals that fall into the accumulating sediment may become encased and produce fossils. With the exception of rare situations, sedimentary rocks are the only type of rock to contain fossils.

Chemical sedimentary rocks are made when dissolved minerals precipitate out of water. For this reason, chemical rocks are sometimes referred to as **precipitates**. Highly soluble minerals such as calcite and halite commonly precipitate to make limestone and rock salt. Because these rocks are often left behind after the water evaporates, they can also be called **evaporites.**

Biologically formed sedimentary rocks are formed as a result of some action of living organisms. It will usually happen from the accumulation of organic matter. Two prime examples are coal and coquina. **Coal** is formed in areas where lush vegetation continually piles up, such as a swamp or bog. The leaves, twigs, and roots are compressed while the plant matter decays. The litter slowly turns into peat—a dry and fibrous mass of twisted vegetation. The peat continues to dry and compress forming lignite. As the process continues, the lignite is turned into the sedimentary rock bituminous coal. If the process continues from here by adding more pressure and heat, the rock will turn into the metamorphic form of coal: anthracite.

Coquina is formed by the accumulation of sea shells. It is a **bioclastic** process, which means fragments (*clastic*) of life (*bio*). The shells, made mostly of calcite, collect in a large group and get buried by further deposition. While being subjected to the pressure of burial, the calcite in the shells will simultaneously dissolve and precipitate—essentially "melting" the shell fragments together. This is similar to adding water to a cup of ice and the ice cubes melting together.

Lesson 3-4 Review

1. Define clastic.
2. Define bioclastic.
3. Describe how land-derived sedimentary rocks form.
4. What is the difference between breccia and conglomerate?
5. What type of rock is made of particles larger than .005 cm?
   a) shale b) siltstone c) sandstone d) conglomerate

6. What type of rock is made of particles .1 cm?
   a) shale b) siltstone c) sandstone d) conglomerate
7. What is the range of grain sizes in siltstone?
   a) less than .0004 cm
   b) .0004 cm–.006 cm
   c) .006 cm–.2 cm
   d) mixed
8. What are evaporites?
   a) rocks made from dissolved mineral left behind after the water evaporates
   b) rocks made from the accumulation of sediments of mixed sizes
   c) rocks made from sea shells cemented together
   d) the original chemicals that eventually form intrusive rocks

Lesson 3–5: Metamorphic Rocks

Metamorphic rocks are rocks that have "changed form" from some other rock. Simply put, if you have a rock, any rock at all, and add enough heat, the atoms will rearrange somehow and make a new rock. The original, or "parent," rock can be igneous, sedimentary, or another metamorphic rock. The two categories of metamorphic rocks are regional and contact metamorphic rocks.

Contact metamorphic rocks, as the name implies, are formed when a preexisting rock comes in *contact* with extreme heat. The heat will come in the form of being touched by melted rock. When magma touches some other rock, the rock will get "fried" and change some of its properties. The atoms will rearrange, or recrystallize, which in turn changes the properties of the original minerals. This is similar to what happens when you cook. When the food comes in contact with the hot pan, it changes some of the food's properties.

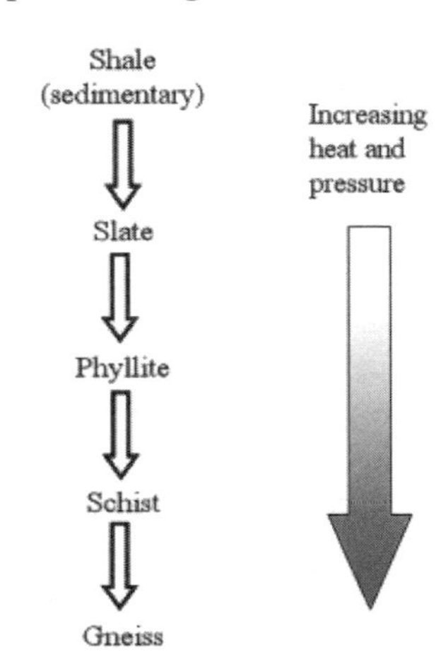

Figure 3.8
Metamorphic Rocks

Regional metamorphic rocks are changed by the combination of heat and pressure working together. Regional metamorphism gets its name from the fact that this process will usually happen over wide regions. An excellent environment for regional metamorphism is where continents are colliding creating mountains. Under the extreme heat and pressure, rocks can flow and bend, and the atoms can rearrange. These types of rocks will look warped and distorted. Instead of flat sedimentary layers, the metamorphic version will have kinked and bent layers.

Metamorphic rocks that can be identified by their patterns of warping are called **foliated**. Slate, which is only a slightly metamorphosed version of shale, has thin, flat layers, which split easily. Phyllite is slate that has undergone more metamorphism. The added pressure has squished some of the minerals flat forming tiny mica flakes. The result is essentially a piece of slate that is shiny. Take phyllite and add more heat and pressure and you'll get the shiny and sparkly schist. Schist has mica flakes large enough to see. The final stage of this development is the formation of gneiss. In gneiss, the minerals have rearranged into bands that often resemble zebra stripes.

Non-foliated metamorphic rocks are a little more difficult to identify because they lack the tell-tale patterns that help identify the foliated rocks. Non-foliated rocks are usually identified by their composition. Two great examples are marble and quartzite. The two rocks can look identical: very light colored, and a freshly broken piece will look like packed snow or ice. However, marble is made mainly of calcite, and quartzite is made from quartz. Calcite is softer on Mohs scale of hardness and will react with diluted hydrochloric acid. The quartzite is very hard and does not react with acid.

Type of Metamorphism	Parent Rock	Metamorphic rock	
Regional	Shale	Slate	Foliated
	Slate	Phyllite	
	Phyllite	Schist	
	Schist	Gneiss	
Regional or Contact	Sandstone	Quartzite	Non-foliated
	Limestone	Marble	
	Conglomerate	Metaconglomerate	
Contact	source rocks vary	Hornfels	Non-foliated

Lesson 3–5 Review

1. List the two types of metamorphism.
2. How is foliated different from non-foliated?
3. Name three types of foliated metamorphic rocks.
4. What is the parent rock of slate?
   a) phyllite b) schist c) quartzite d) shale
5. Which metamorphic rock is formed from the metamorphism of sandstone?
   a) phyllite b) schist c) quartzite d) shale
6. Which rock has experienced the most heat and pressure?
   a) slate b) gneiss c) hornfels d) schist
7. What process forms hornfels?
   a) heat c) heat and pressure
   b) compaction and cementation d) pressure

To Sum up Rocks

Sedimentary:

- Clastic rocks are identified by texture:
  - Sand = sandstone
  - Silt = siltstone
  - Clay = shale
  - Mixed = conglomerate (rounded sediments) or breccia (angular sediments)
- Non-clastic sedimentary rocks are identified by their composition such as halite (rock salt).

Igneous:

- Identified by a combination of color (light or dark) and texture (glassy, fine, or coarse). Color matches up with composition: Light is a felsic composition, dark is a mafic composition.

Metamorphic:

- Foliated rocks have a texture that helps identify them.

Chapter Exam

Exercise A

Write igneous, sedimentary, or metamorphic on the line next to each descriptive phrase.

1. ______________ rocks are formed from cooling and hardening of magma.
2. ______________ rocks are formed by compaction and cementation.
3. ______________ rocks may be foliated or non-foliated.
4. ______________ rocks are formed by changes in existing rock.
5. ______________ rocks may be classified as clastic, bioclastic, or chemical.

Exercise B

Decide whether each statement is true or false. In the space provided write T or F. If the statement is false; correct the underlined word or words.

_______6. Scientists believe that when the Earth was first formed, its crust was made of <u>metamorphic</u> rock.

_______7. When magma cools slowly, the minerals within it form <u>large</u> crystals.

_______8. When different minerals in one rock have formed bands or layers, the rock is classified as <u>clastic</u>.

_______9. <u>Nitrogen</u> and <u>carbon dioxide</u> are the two most abundant elements in the crust.

Exercise C

Fill in the blanks with the word or words that will make the sentence true. Use the following words.

foliated	granite	gneiss
chemical	sandstone	size

10. Clastic rock can be subdivided according to the ______________ of the rock particles.
11. An example of igneous rock is ______________ ; an example of sedimentary rock is ______________ ; an example of metamorphic rock is ______________ .

12. Slate and schist are examples of _______________ metamorphic rocks.
13. Rock salt and gypsum, minerals that remain when water has evaporated, are examples of _______________ sedimentary rock.
14. What is the relationship between the rate of cooling of molten rock material and the size of the mineral grains formed?
    a) The faster the rate, the larger the mineral grains.
    b) The slower the rate, the smaller the mineral grains.
    c) The slower the rate, the larger the mineral grains.
    d) There is no direct relationship between rate and mineral grain size.
15. According to the Igneous Rock Chart (pg. 77), basalt is
    a) coarse-grained, extrusive rock.
    b) coarse-grained, intrusive rock.
    c) fine-grained, intrusive rock.
    d) fine-grained, extrusive rock.
16. Metamorphic rock, such as marble, made of only one mineral, is
    a) intrusive.
    b) foliated.
    c) polymínerallic.
    d) monomineralic.
17. Rock formed from particles of other rocks is called
    a) crystallized rock.
    b) clastic rock.
    c) extrusive rock.
    d) foliated rock.

Answer Key

Answers Explained Lesson 3–1

1. **A** is incorrect.

   B is **correct** because all physical properties, including hardness, comes from the internal arrangement of atoms.

   C is incorrect.

   D is incorrect because fracturing or cleaving are different characteristics than hardness. Hardness tests resistance to scratching; fracture and cleavage describe ways that the mineral will break.

2. **A** is incorrect because white is one of the most common mineral colors.

   B is **correct** because having a white sample of a mineral does little to help identify which mineral it is.

C is incorrect.

D is incorrect because there are many other white minerals besides quartz.

3. **A** is incorrect because breaking along three planes that are each at 90° from each other is cubic cleavage.

   B is incorrect because imperfect cleavage can be any type (basal, cubic, rhombohedral) but does not produce completely flat planes.

   C is **correct** because basal cleavage splits into flat sheets.

   D is incorrect because feldspar is a mineral, not a type of cleavage.

4. **A** is **correct** because a hardness test involves two minerals: the unknown and a known mineral.

   B is incorrect because hardness is resistance to scratching.

   C is incorrect because hitting with a hammer will test a mineral's brittleness or malleability.

   D is incorrect because splitting it into thin sheets would demonstrate a mineral's cleavage.

5. **A** is incorrect.

   B is **correct** because calcite will fizz or effervesce in dilute hydrochloric acid.

   C is incorrect. Although gold will dissolve in acid the test is not needed because most metals will react with acid.

   D is incorrect because hardness is resistance to scratching—a physical property. Reaction to acid is a chemical property.

6. **A** is incorrect because a glassy luster in a type of non-metallic luster.

   B is incorrect because a resinous (also known as waxy) luster will often look oily.

   C is incorrect because a waxy (also known as resinous) luster will often look oily.

   D is **correct** because glassy is the same as vitreous.

Answers Explained Lesson 3–2

1. As magma cools, it forms **igneous** rock by the process of **solidification** .
2. Igneous rocks can form **igneous** , **sedimentary,** and **metamorphic** rocks.
3. Sediments form **sedimentary rocks** by the processes of **compaction** and **cementation** .
4. Sediments form from the process of **weathering** .
5. Sedimentary rocks can form **igneous** , **sedimentary,** and **metamorphic** rocks.

6. **Exposure to extreme heat and/or pressure will change igneous rock into metamorphic rock.**
7. **Sedimentary rocks can be changed into igneous rock by melting and then solidification.**
8. **Metamorphic rock will change into sedimentary rock after weathering, erosion, burial of sediments, and compaction and cementation.**
9. Metamorphism involves the addition of **heat** and **pressure** to preexisting rocks.
10. Compaction and cementation of sediments forms **sedimentary** rocks.
11. Subjecting sedimentary rocks to extreme heat and pressure forms **metamorphic** rocks.
12. Solidification of molten minerals forms **igneous** rocks.

Answers Explained Lesson 3–3

Exercise A

Rock	Formed from	Cooling rate	Crystal size	Texture
Rhyolite	Lava	**Fast**	Small	**Fine**
Gabbro	**Magma**	Slow	**Large**	Coarse
Basalt	**Lava**	Fast	**Small**	Fine
Pumice	Lava	**Very Fast**	No crystals	**Glassy**
Obsidian	**Lava**	Very fast	**No crystals**	Glassy
Granite	Magma	**Slow**	Large	**Coarse**

Exercise B

Olivine	**4** %
Pyroxene	**37** %
Plagioclase feldspar	**52** %
Biotite mica	**0** %
Hornblende (Amphiboles)	**6** %

Multiple Choice

1. Granite is **intrusive**. It has large crystals and a coarse texture.
2. **A is correct**. Plutonic is also referred to as **intrusive** because a pluton is an internal magma structure.
3. **A glassy texture has no crystals; a fine texture has small crystals. A glassy texture was cooled very quickly—too fast for any crystals to form.**

4. Andesite contains: **orthoclase feldspar**
quartz
plagioclase feldspar
mica
hornblende

The most mafic types of andesite contain a little pyroxene.

5. a) **Rhyolite** has a lower density because it is further to the left on the Igneous Rock Chart.
b) **Rhyolite** has a lighter color because it is further to the left on the Igneous Rock Chart.
c) **Basalt** has a more mafic composition because it is further to the right on the Igneous Rock Chart.

6. **Granite** is felsic (on the left side of the chart) and intrusive (near the top of the chart).

7. Felsic igneous rocks generally have a **lighter** color.

8. **A** is incorrect because intrusive rocks can be either light-colored (granite) or dark-colored (gabbro).
B is **correct** because intrusive rocks cool slower than extrusive rocks, which gives the crystals more time to grow.
C is incorrect because intrusive rocks are coarser than extrusive rocks because the take longer to cool.
D is incorrect because intrusive rocks tend to be denser than extrusive rocks although there are exceptions.

9. **C** is **correct** because granite has coarse crystals that are between 1 mm and 10 mm.
A, B and **C** are incorrect.

10. **C** is **correct** because granite has a coarse texture which is between 1 mm and 10 mm.
A, B and **D** are incorrect.

Answers Explained Lesson 3–4

1. Clastic means to come from broken fragments of rock. This will produce sediments.

2. Bioclastic means "fragments of life." It is the accumulation of organic remains such as shells, bones or dead plants.

3. A solid rock (igneous, sedimentary, or metamorphic) is exposed to weathering, which reduces the rock into sediments. The sediments accumulate and thicken. While under pressure from the sediment pile above, mineral cement in the groundwater will glue the particles together making it a solid rock.

4. Conglomerate and breccia are both clastic sedimentary rocks made from mixed sizes of sediments. The difference between the two rocks is that conglomerate is made from rounded sediments that have been weathered more, breccia has been made from more angular sediments that were freshly weathered just before the sediments accumulated into the breccia.

5. **A** is incorrect because shale is made from sediments that are less than .0004 cm, which is smaller than .005 cm.

   B is **correct** because .005 cm is in between .0004 cm–.006 cm which is the size of silt.

   C is incorrect because the sand to make sandstone has a range of .006 cm–.2 cm, which is bigger than .005 cm.

   D is incorrect because conglomerate is made from varied sizes of sediments.

6. **A** is incorrect because clay particles, which make shale, are too small: .0004 cm–.006 cm.

   B is incorrect because silt particles, which make siltstone, are too small: less than .0004 cm.

   C is **correct** because .1cm is within the range for sand sandstone: . 006 cm–.2 cm

   D is incorrect because conglomerate is made from varied sizes of sediments.

7. **A** is incorrect because less than .0004 cm is the size range for clay particles that will form shale.

   B is **correct** because siltstone is made from silt, which has a particle size range from .0004 cm to .006 cm

   C is incorrect because the size range: .006 cm–.2 cm is that of sand that will make sandstone.

   D is incorrect because mixed particles will make either conglomerate or breccia.

8. **A** is **correct**. Evaporates are made when water dissolves minerals. The water later evaporates and leaves the dissolved minerals behind as a deposit of rock material.

   B, **C**, and **D** are incorrect.

Answers Explained Lesson 3–5

1. Regional metamorphism and contact metamorphism
2. Foliation will have some kind of pattern in the rock such as stripes or bands; non-foliation will not show the patterns.
3. Slate, phyllite, schist, and gneiss are all foliated metamorphic rocks.

4. **A** is incorrect because phyllite will make schist if it undergoes further regional metamorphism.

   B is incorrect because schist will make gneiss if it undergoes further regional metamorphism.

   C is incorrect because quartzite will continue to be quartzite if it undergoes further regional metamorphism.

   D is **correct** because shale will turn into slate if it undergoes regional metamorphism.

5. **A** is incorrect because phyllite is made from the metamorphism of slate.

   B is incorrect because schist is made from the metamorphism of phyllite.

   C is **correct** because sandstone, which is made from sand is mainly quartz, will turn into quartzite after metamorphism.

   D is incorrect because shale is a sedimentary rock.

6. **A** is incorrect because slate has experienced only a little heat and pressure.

   B is **correct** because gneiss is the metamorphic product of extreme heat and pressure.

   C is incorrect because hornfels is made from heat only.

   D is incorrect because schist has not undergone as much heat and pressure as gneiss.

7. **A** is **correct** because hornfels is made by contact metamorphism, which is only the exposure to heat.

   B is incorrect because compaction and cementation will form a sedimentary rock.

   C is incorrect because hornfels is not a regional metamorphic rock and is not formed by the combination of heat and pressure.

   D is incorrect because pressure is not involved in the formation of hornfels.

Answers Explained Chapter Exam

Exercise A

1. **Igneous** rocks are made from the cooling of magma.
2. **Sedimentary** rocks are made from sediments that are compacted and cemented into solid rock.
3. **Metamorphic** rocks are divided into foliated and non-foliated types.
4. **Metamorphic** rocks are rocks that have "changed form."
5. **Sedimentary** rocks may be classified as clastic, bioclastic, or chemical.

Exercise B

6. F igneous
7. T
8. F foliated
9. F Silicon, oxygen

Exercise C

10. **Size**. The difference between shale, siltstone, sandstone, conglomerate, and breccia is the size of the particles within them.
11. Igneous rock: **granite**
    Sedimentary rock: **sandstone**
    Metamorphic rock: **gneiss**
12. **Foliated**. Slate and schist have layers and flakes.
13. **Chemical**. Evaporites are formed by a chemical process.
14. **A** is incorrect because the faster the rate, less time the crystals have to grow.
    B is incorrect because slower cooling rate gives larger crystals.
    C is **correct.**
    D is incorrect. The relationship is an inverse relationship: The faster the cooling rate, the smaller the crystals.
15. **C** is correct.
16. **A** is incorrect because *intrusive* is a term for igneous rocks formed inside the Earth.
    B is incorrect because *foliated* is a term for metamorphic rocks that show layers, banding, or flakes.
    C is incorrect because polymineralic means to be made of many minerals mixed together.
    D is **correct.**
17. **A** is incorrect because crystallized rock is made of crystals minerals, not particles (or sediments) of rock.
    B is **correct** because clastic means "fragments" of other rocks.
    C is incorrect because extrusive rock is a type of igneous rock made from cooled lava.
    D is incorrect. Foliated rock is a type of metamorphic rock that has layers, bands, or flakes. It may have been formed at one time from the particles of other rocks, but it also may have had a different origin. B is a better answer because it is *always* true.

Earthquakes and Volcanoes

Lesson 4–1: Layers of the Planet

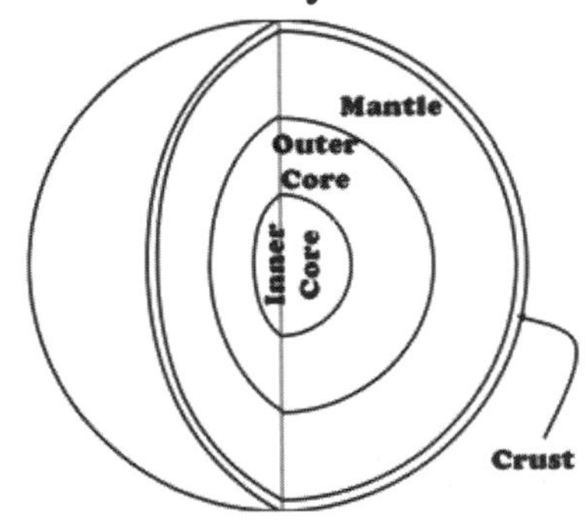

Figure 4.1 Cross section

When the Earth was first forming, it was a ball of molten rock held together by gravity. Under this gravitational influence, the heavier, denser materials "sank" towards the center and the lighter material migrated towards the surface. Eventually, this migration created several zones within the planet with different minerals and characteristics.

Crust

Starting at the surface and working our way down, the first layer that we reach is the crust. The **crust** is the layer on which we live and from which we get all of our natural resources. Despite this important distinction, the crust is the thinnest of all the layers. Thickest under continental mountains and thinnest at the cracks in the ocean floor, the crust has a range of thicknesses from 2 km to 12 km. In fact, the crust is so thin compared to the size of the Earth that page-sized true-scale drawings have difficulty drawing it thin enough. Often the ink of the line is too wide to show the correct size! The entire crust of the Earth can be thought of as "floating" on the liquid inner layers deep below. In fact, floating is an appropriate description because the solid crust is less dense than the liquid rock and is supported by this difference in density just as a sheet of ice on the ocean is. The crust, comprised of the continents and the ocean bottoms, has properties that contrast so vividly that it is considered to have two different types of crust material: continental crust and oceanic crust.

The **continental crust** is thick, yet low in density: 3.7 g/cm^3. The low density and lighter color of the continents are attributed to the mainly granitic minerals in the continental crust. Because it has the lowest density of all the layers, it floats the highest and, therefore, is the part that "sticks up."

The **oceanic crust** is made of darker, denser (5.0 g/cm^3) basalt. The basaltic crust is formed mostly from mid-ocean ridges where submarine volcanoes spew out lava that forms new crust. The basalt, being denser than granite, floats lower on the surface of the mantle than the continental crust. When water falls on the surface of the Earth (rain), it collects into the lowest parts and fills in the depressions between the higher continents. If the world had no water, there would still be a stark difference between the two types of crust due to the thicknesses of the two crusts.

Mantle

Below the crust is the mantle. The **mantle** is a very thick layer made primarily of the mineral olivine stretching from the bottom of the crust to the outer core. Due to the lower pressure (relatively speaking), the upper portion of the mantle has a lower melting point and is very close to melting. Actually, because the mantle is made of many minerals, each with their own melting points, some of the minerals are melted. This upper part of the mantle is sometimes referred to as its own layer called the **asthenosphere**.

The asthenosphere is a partly molten layer on which the crust floats. It has a gooey consistency kind of like peanut butter. Because it is in between solid and liquid, the asthenosphere can flow very slowly and has convection currents. It is the convection currents within the asthenosphere that cause the continents to drift around the surface of the planet slowly over millions of years.

The bottom of the asthenosphere is a zone where the properties of the mantle change from partly molten to the more rigid material of the mantle proper. This interface was discovered by a seismologist named Mohorovičić and the feature is named after him. Because of the difficulty in spelling, punctuating, and pronouncing his name correctly, the zone is often referred to as the **Moho**. Above the Moho, the properties of the asthenosphere (mainly seismic wave speeds) are stable. Below the Moho, in the mantle, the properties are also stable. However, right at the Moho, the properties drastically change from those of the asthenosphere to those of the mantle. For this reason, the Moho is sometimes called the **Moho discontinuity**—the properties are not continuous across the zone.

Deeper, within the remainder of the mantle, the material is still very hot and the rock still has convection—but at a much slower rate than the asthenosphere.

Core

Below the mantle, we see the **core** of the Earth, which is divided into two sections: the **outer** and the **inner core**. Strangely enough, the outer core is molten and the inner core is solid. During Earth's formation, some of the densest material migrated to the core. The metals iron (Fe) and nickel (Ni) make up the vast majority of the two core layers. As a result, the core is sometimes referred to as having a "FeNi" composition. The outer core has a density of 10 g/cm^3 while the highly pressurized inner core has a density of 13 g/cm^3.

Why Is the Outer Core Liquid While the Inner Core is Solid?

At first thought this seems like a strange arrangement. However, there is a good reason for this. Once you penetrate the crust of the Earth and travel towards the center of the planet, the pressure and temperature both increase. The melting point of any material depends on pressure. Think of it this way: The loose atomic arrangement of a liquid needs more room for the atoms to move freely around. If it is under so much pressure that the atoms cannot move around, then they will stay in the more compact, solid phase. At the interface between the inner and outer cores, the pressure becomes too great for the iron and nickel to liquefy and even though it is hotter than the outer core it cannot melt. If Captain Kirk's *Enterprise* used the Star Trek transporter to beam the inner core out into space, where it suddenly had no pressure constraining it, the core would instantly expand and turn to a liquid!

Lesson 4-1 Review

1. Name the layers of Earth in order from densest to lightest.

____

____

____

2. Which of the following is liquid?
   a) oceanic crust
   b) outer core
   c) inner core
   d) continental crust

3. Sometimes the Earth's core is referred to as "FeNi." Where does this term come from?
   a) FeNi is a combination of the names of the two scientists that discovered the core.
   b) The core is made from iron and nickel.
   c) FeNi comes from the French term *fini*, which means "the end" or in this case, the center.
   d) FeNi is a term referring to the magnetic properties of the core.

4. Through most of the Earth, the properties are very consistent or, change gradually. Where do the properties change drastically?
   a) at the Moho
   b) in the mantle
   c) in the outer core
   d) in the inner core

5. Continental crust is mostly made of
   a) granite. b) basalt. c) magma. d) lava.

6. Oceanic crust is mostly made of
   a) granite. b) basalt. c) magma. d) lava.

7. Continental crust is
   a) thinner and denser than oceanic crust.
   b) thicker and denser than oceanic crust.
   c) thinner and lighter than oceanic crust.
   d) thicker and lighter than oceanic crust.

Lesson 4–2: Movement of the Crust

Earthquakes, volcanoes, and mountain ranges occur around the planet in many of the same areas and seem to be related. The Earth's crust is broken into many pieces called **plates**. It gives the world the appearance of an egg that has been broken and re-assembled. Earthquakes, volcanoes, and mountain ranges, otherwise known as **crustal activity**, are concentrated along the edges of these plates. The reason is simple: The edges of the plates are where the sections of the crust grind against each other or get torn apart. The centers of the plates (with a few exceptions mentioned a little later) tend not to have much in the way of activity.

Most of the world's crustal activity is concentrated along the shores of the Pacific Ocean. At any place along the Pacific Rim, you are not far from mountain chains (or island chains), active earthquake areas, or volcanoes. Because of the concentration of earthquakes and volcanoes, the Pacific Rim has earned the nickname "**The Ring of Fire**."

The exceptions to this concentration of seismic activity along the edges of plates are "hot spots." **Hot spots** are plumes of hot mantle material that melt through the middle of plates (rather than belts along the edges). These spots stay stationary while the plate moves above it. The spot melts through the plate like a blow torch and produces a volcano above it. As the plate moves, the spot melts through another spot-producing a chain of volcanic islands. Hawaii is an example of a hot spot island chain.

Types of Volcanoes

A **cinder cone** volcano is one that is made, as the name suggests, of cinders. It is made when loose rock material that has been ejected piles up around the central vent of the volcano. These cones can be steep but erode quickly.

A **shield volcano** is formed by lava flows. Because it is made from the accumulation of solidified liquid, these volcanoes tend to be very wide with gentle slopes. Near the peak of a shield volcano is a large crater called a caldera. The **caldera** is made of several smaller vents, some of which may have active lava flows.

A **composite cone** is made from alternating lava flows and deposits of cinders. The cinders allow the cone to get quite steep and the lava flows hold the cone together. A composite cone has the "typical" shape envisioned when people think of what a volcano looks like.

Lesson 4–2 Review

1. Most geologic activity occurs
   a) near populated areas.
   b) in the centers of the continents.
   c) along the edges of the plates.
   d) in Antarctica.

2. Chains of mountains, earthquakes, and volcanoes can be described as
   a) being randomly scattered around the planet.
   b) rare events in modern times.
   c) the main causes of the weather.
   d) related events that happen in similar areas.

3. Hawaii is an exception to where we typically find active volcanoes because
   a) Hawaii is a populated area.
   b) Hawaii is in the center of a plate.
   c) volcanoes can't erupt in water.
   d) Hawaii is on the edge of a plate.

4. Because of the frequency of activity around the rim of the Pacific Ocean, it has the nickname
   a) the Ring of Fire.
   b) the Ring of Doom.
   c) a hot spot.
   d) Dante's Peak.

Lesson 4–3: Plate Boundaries

The places where plates meet and interact, called **plate boundaries**, fall into three types depending on their relative movements: convergent, divergent, and transform.

Convergent Plate Boundaries

A **convergent plate boundary** is an area where two plates are colliding or converging into each other. This collision zone will have a few different appearances depending on the combination of the two plates colliding: continental or oceanic.

Continental-Continental

In a **continent-continent collision** two pieces of continental crust slam into each other. Both plates are thick and have a low density. As a result, they buckle during the collision. This produces very high mountains as well as deep "roots" to the mountains where the collision buckles downward. This is where you would find the absolute thickest sections of the crust on the planet. The Himalaya Mountains are a prime example of a continent-continent collision, where India is colliding into southern Asia. The collision is still occurring today and the mountains, including the world's highest, Mount Everest and K2, are still growing.

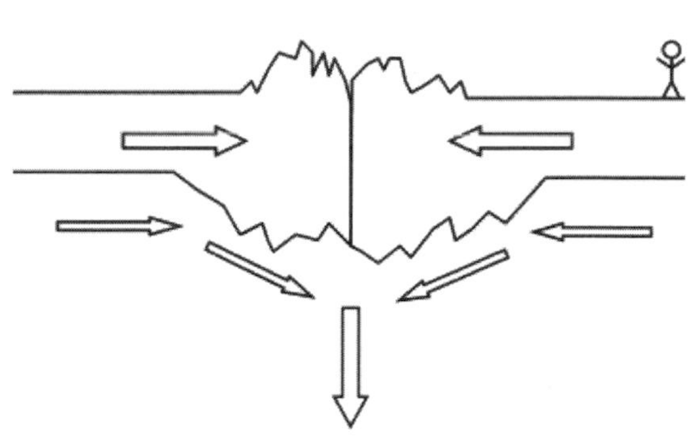

Figure 4.2 Continental-Continental

Continental-Oceanic

When a continental plate collides into an oceanic plate there is a sharp contrast in the density of the plates. The denser oceanic plate is forced under the lighter continental plate in a process called **subduction**. The continental plate is forced upward, forming mountains. Meanwhile, the oceanic plate sinks down bringing along a section of the ocean floor with it, creating a deep trench. As the oceanic plate subducts it dives deeper into the hot mantle, where it eventually melts and is destroyed. The progress of the subducting plate can be traced by monitoring the source of earthquakes. The quakes occur where the two plates grind against each other and can be traced along the edge of the diving oceanic plate. The friction of the grinding will also cause melting of the

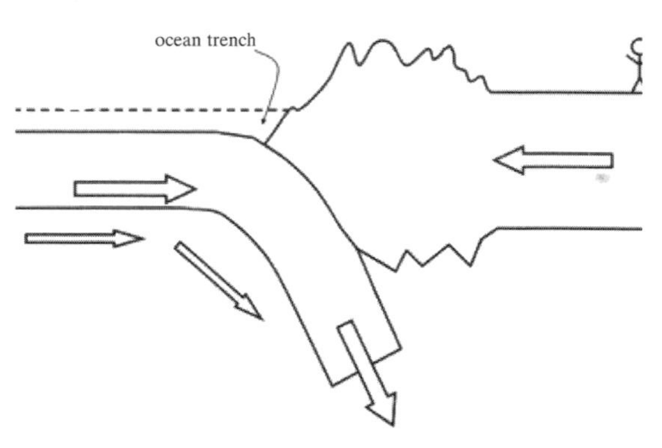

Figure 4.3 Continental-Oceanic

rock material. The molten rock, being less dense than the surrounding rock, migrates upwards and produces volcanoes scattered throughout the mountain chain above.

A good example of a continental-oceanic collision is the west coast of South America. On the continent side of the collision the Andes Mountains are forming. Just off the coast, on the oceanic side, there is a deep ocean trench.

Oceanic-Oceanic

In an oceanic-oceanic collision two thin, dense plates collide. Because both plates are dense and tend to sink, this type of collision produces a deep ocean trench. Even though both plates are dense, one plate will often sink under the other. as with the ocean-continent collision, the friction of the grinding plates causes melting. However, in this case because there is no mountain chain caused by buckling, the lava produces a chain of volcanic islands that dot the edge of the overriding plate. The western side of the Pacific Ocean is littered with examples of ocean-ocean colliding plates. The Philippine Archipelago is a chain of volcanic islands and just off the coast the deepest trench in the world, the Marianas Trench, dives 11km down.

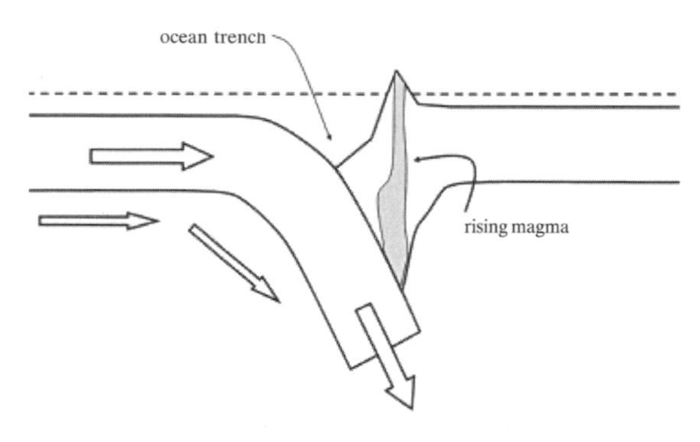

Figure 4.4 Oceanic-oceanic

Divergent Plate Boundaries

A **divergent plate boundary** is caused by two plates splitting apart. Most of the world's current divergent plates are in the oceans, although there are some that have happened on continents. When two plates split, it creates a crack through the crust through which lava will well up. This creates new crust as the plates move apart. The upwelling of fresh lava makes a low, underwater mountain chain called a ridge along the edges of the two plates.

One of the most well-known divergent plate boundaries is the Mid-Atlantic Ridge. This ridge extends from the northernmost Atlantic Ocean down to the South Atlantic between the tips of Africa and South America. The Mid-Atlantic Ridge formed when Pangaea split apart separating Europe and Africa from the Americas and has been splitting ever since. The lava that comes out of the ridge system is basalt forming new oceanic plate material except in Iceland. Iceland is a section of the Mid-Atlantic Ridge that has surfaced above sea level. This is because instead of spewing out basaltic rock, that section is releasing granitic continental rock.

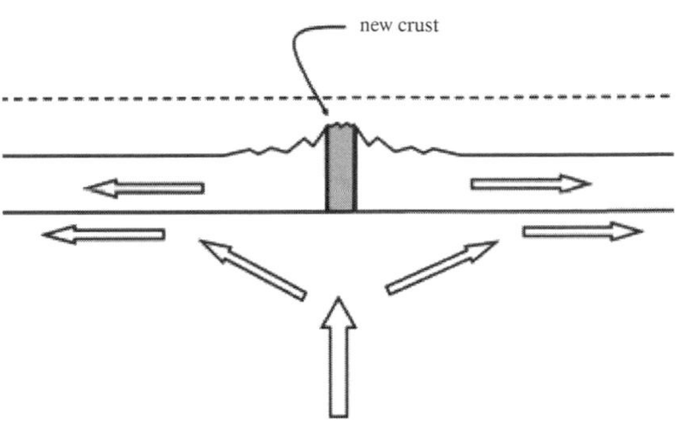

Figure 4.5 Divergent

There are several divergent mid-ocean ridges around the world. However, there are a few places where continents have divergent plate boundaries. The long valley that is home to the Nile River in Africa is called the Great Rift Valley. This is the site of a failed rifting millions of years ago. Forces within the Earth attempted to pull Africa apart but they just ran out of steam. The area was pulled apart enough to start stretching the crust so that it became a little thinner creating the valley. Not far away, the Arabian Peninsula separated from mainland Africa creating the Red Sea.

Transform Plate Boundaries

A **transform plate boundary** is caused by two plates sliding past each other. While they pass each other they grind, sometimes with extreme results. Technically speaking, transform plates slide laterally past each other, which means "to the side" (just like a lateral pass in football). The effects on the surface are offset features such as roads, fences, and rivers. Probably, the most famous and studied transform plate boundary is the San Andreas Fault in California.

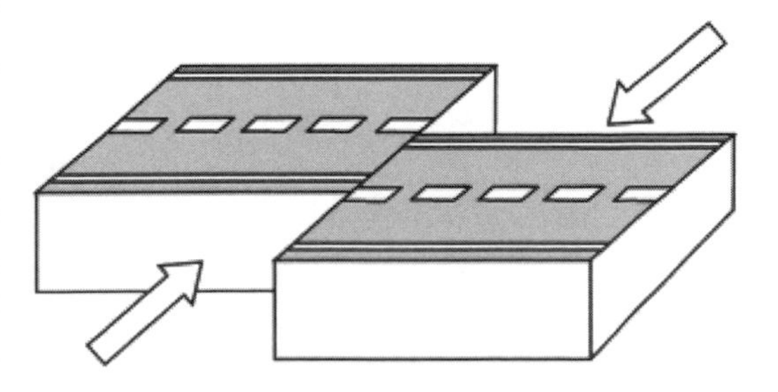

Figure 4.6: Transform

Along the San Andreas Fault there are many cases of fences, roads, and rivers that at one time were straight and are now "zig-zagged" in a stair-stepped fashion. This is because the west side of the fault (which is on the edge of the Pacific Plate) is moving northwest while the east side of the fault is moving southeast (relatively against each other).

In fact, both plates are moving in just about the same direction—the Pacific Plate is just moving faster to the northwest. There is some truth to the myth that someday California will break off of the mainland—given a few dozen thousand years! Eventually, the small motions along the San Andreas Fault will slowly move all of California to the west of the fault off of the coast.

The motions of the plates are caused by convection below the crust within the asthenosphere. The study of the currents inside the Earth and their affect on the overlying plates is **plate tectonics**. In areas where the thick semimelted rock is hotter, it rises. When the rising material reaches to crust, it spreads to the sides, taking the plates above with it. As a result, divergent plate boundaries are located above areas of upwelling in the asthenosphere. In areas where the asthenosphere is a little cooler, it sinks pulling the surrounding material inward and downward. On the surface, this will create convergent plate boundaries. In the places where convection cells don't line up directly into each other or away from each other, transform plate boundaries will mark the location of currents sliding past each other.

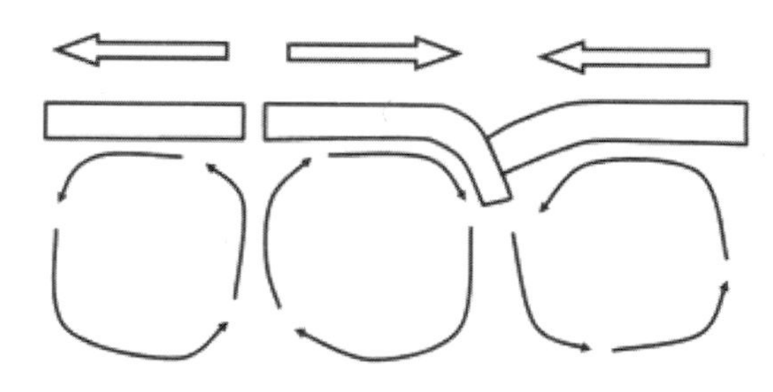

Figure 4.7

Lesson 4-3 Review

1. What does an ocean plate do after it collides with a continental plate?
   a) The ocean plate crumples up and forms underwater mountains.
   b) The ocean plate slides on top of the continental plate.
   c) The ocean plate sinks below the continental plate.
   d) Both plates sink.

2. Which type of plate is denser?
   a) continental
   b) oceanic
   c) They are the same.
   d) transform

3. At which type of plate boundary is new crust formed?
   a) convergent plate boundary
   b) divergent plate boundary
   c) transform plate boundary
   d) all of the above

4. A trench is associated with which type of plate boundary?
   a) convergent plate boundary
   b) divergent plate boundary
   c) transform plate boundary
   d) all of the above

5. Why does an ocean plate subduct under a continental plate when they collide?
   a) The ocean plate is denser than the continental plate.
   b) The continental plate is denser than the oceanic plate.
   c) The mountains on the continental plate hold down the ocean.
   d) There is no reason; it is random.

6. Which best describes a transform plate boundary?
   a) a boundary between plates that are moving apart
   b) a boundary between plates that are moving horizontally past one another
   c) a boundary between plates that are moving toward each other
   d) a boundary between plates where new crust is created

7. When one crustal plate plunges down under another, it is referred to as
   a) isostatic rebound.
   b) seafloor spreading.
   c) subduction.
   d) rifting.

8. Which choice best describes the Mid-Atlantic Ridge?
   a) This is a zone of little crustal activity.
   b) This is a subduction zone.
   c) One crustal plate is plunging beneath another.
   d) New crust is forming where seafloor spreading occurs.

Lesson 4-4: Plate Tectonics

Continental Drift

There is some strong evidence that the plates and continents of the world are slowly moving around, but the concept of **continental drift** is still considered a theory. As with any good theory, there are several pieces of supporting evidence to validate the theory.

One of the first clues to the existence of continental drift was discovered by a scientist named Alfred Wegener. Shortly after a good portion of the world was mapped, Wegener noticed that some of the continents had shorelines that seemed to have shapes that would fit together. He cut up a map of the world and realized that most of the world's landmasses could be pieced together much like a jigsaw puzzle. Specifically, the west coast of Africa and the east coast of South America fit together perfectly.

Wegener recreated an ancient landmass made of most of the world's current continents, which is now referred to as "**Pangaea**." When the supercontinent of Pangaea was together, you would be able to walk, without crossing anything more than rivers, from California to China and Russia down to Argentina. Pangaea was a solid landmass from approximately 260 million years ago until about 100 million years ago. Pangaea began to split with the birth of the Atlantic Ocean, which separated the Western Hemisphere continents from the Eastern.

Fossil Evidence

An interesting "side effect" of Pangaea being together once and now separated is the history that is recorded with fossils. While Pangaea existed, land animals and plants were free to migrate easily across the land. Plants can migrate whenever their seeds travel in the wind or are carried by animals. When Pangaea broke apart, the new continents carried the fossils of these organisms along with them. We now find fossils of the same creatures on opposite shores. The simplest explanation for this is that the organisms lived over wide areas of the ancient supercontinent while it was connected.

Similar to matching fossils on opposite shores, unique mineral deposits also line up when the continents are reassembled. Likewise, mountain chains such as the Appalachians continue "across the Atlantic Ocean" off the coast of Canada and into northern Europe.

Coal in Antarctica

The existence of coal in Antarctica also fits nicely into the theory that the continents have changed their positions. Coal is formed where large amounts of vegetation accumulate over very long periods of time. The coal in Antarctica was formed in tropical swamps when the continent was much closer to the equator. It has since moved, along with its coal, to its present location at the South Pole.

Glaciers

When glaciers crawl over the surface of the Earth, they make unmistakable changes in the actual shape of the land and leave their calling cards behind. One type of calling card left behind is an indication of "direction of travel." Glaciers leave very clear marks behind that tell us exactly which direction they were moving. For more details on these features, see the section on glaciers in Chapter 5.

Particularly in the Southern Hemisphere, the glacial evidence left behind what seems like several random ice flows—one of which is in India coming from the south! However, when Pangaea is reassembled, the picture of one or two large ice sheets comes together. India, which was attached to Antarctica at the time, now has a consistent story to tell.

Magnetic Stripes

This piece of evidence is very compelling but requires a couple of background points:

Point 1: Lava is a mixture of melted minerals of which Iron is a very common one. Iron is attracted to the Earth's magnetic North Pole and prefers to line up with Earth's magnetic field. While lava is liquid, the bits of iron will tend to line up with a magnetic orientation pointing at the North Pole. When the lava solidifies, the magnetic orientation is frozen into the rock—essentially; it will have a "north end" and a "south end."

Point 2: During World War II, the United States, in an effort to find any advantage that would help out submarines in the war effort, collected any type of data possible about the ocean and ocean floor. When they dragged a magnetic detector along the ocean floor, they discovered that the rocks making up the ocean bottom either had a magnetic orientation pointing to the North Pole or to the South Pole. As the magnetic detectors were dragged along the ocean floor, the orientation of the rock flipped from north to south—but never any other direction. They discovered a very quirky behavior of our planet: For some still-unexplained reason, at random times in Earth's past, the magnetic orientation of the planet reversed—several times. In other words, if you traveled back in time with a compass, sometimes it will point north and other times it will point south.

The significance of these two points becomes clear when we have a map of the magnetic orientation of the entire Atlantic Ocean floor. Imagine that the north-oriented rocks (usually referred to as "normal polarity") are colored white and the south-oriented ("reversed polarity") rocks are colored black. What we will see is a striped pattern that runs parallel to the coasts of the Atlantic shores. In addition, the magnetic stripe pattern that emerges as we cross the Atlantic floor from the United States to the Mid-Atlantic Ridge is an exact mirror-image of the pattern from the Mid-Atlantic Ridge to Europe. This implies that the ocean floor is splitting at the Mid-Atlantic Ridge, new rock is being made at the ridge, and the new rock (along with its magnetic orientation) is split and each half is carried away from the ridge as the ocean floor spreads.

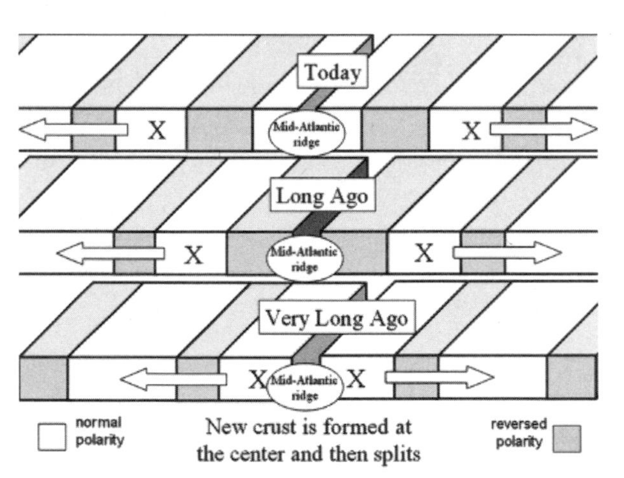

Figure 4.8 Magnetic Stripes

Ages of Continental vs. Oceanic Crust

To add support to the "magnetic stripe" evidence, direct measurements of the ages of the continental rocks and the oceanic rocks also implies that the continents are splitting. The Atlantic sea floor is younger than the continents that made Pangaea. This makes sense because, at the time of Pangaea, the Atlantic had not opened yet and therefore the rocks making up the sea floor had not been made yet. In fact, the oceanic crust all over the planet is younger than the continental crust.

Lesson 4-4 Review

1. What is the name given to the large continent believed to have existed 200 million years ago?
   a) Gondwanaland
   b) Ancestral Siberia
   c) Ancestral North America
   d) Pangaea

2. Which of the following is NOT evidence that the continents were once together?
   a) chains of volcanoes along the Ring of Fire
   b) magnetic stripes on the ocean floor
   c) fossils of the same creatures on opposite sides of the Atlantic Ocean
   d) the "puzzle fit" of the continents

3. Coal is found in Antarctica because
   a) the world was more tilted, which put Antarctica on the equator.
   b) the coal bounced off of Africa during the collision that formed Pangaea.
   c) Antarctica was once in a warmer location.
   d) the heat and pressure of collision formed the coal.

4. From evidence of glaciers in India we can infer that
   a) India was closer to the South Pole at one time.
   b) the world was much colder so that continental glaciers could exist near the equator.
   c) during one of the magnetic reversals, it got very cold in India.

5. The patterns of magnetic stripes on the Atlantic floor
   a) are the same in the Atlantic and Pacific Oceans.
   b) follow lines of latitude.
   c) are lines with matching fossils.
   d) are mirror images on both sides of the Mid-Atlantic Ridge.

6. As you travel from the east coast of the United States to the opposite shore of the Atlantic Ocean, the age of the seafloor bottom will get
   a) older.
   b) younger.
   c) older then younger.
   d) younger then older.

Lesson 4–5: Earthquakes

An earthquake is an event where two pieces of crust shift against each other. The rumbling felt is from the rocks slipping, sticking and breaking. When two sides of a fault rub against each other, there will be friction causing the rocks to stick together. The pressure builds up and the rocks will often deform like a model made of clay. If the pressure and deformation continue, it will get to the breaking point and the rock will suddenly break, causing the earthquake. This slow build-up of pressure and deformation of the rocks is called the "**elastic rebound theory**." This theory explains why the land sometimes changes shape before earthquakes. Roads start to get "s" curves in them, lakes get tilted, and well water changes level.

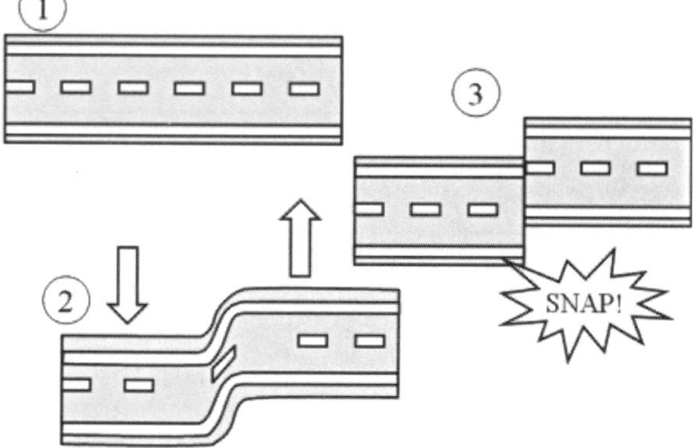

Figure 4.9. When strain builds up on a rock, it will start to deform (2). When it reaches its breaking point (3), the rock will suddenly break, causing an earthquake.

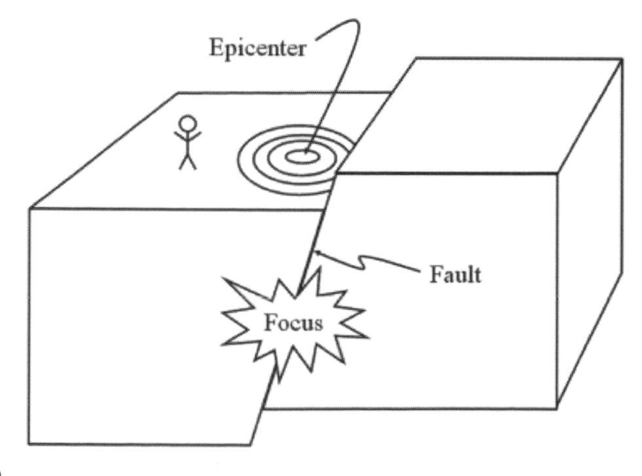

Figure 4.10

The Anatomy of an Earthquake

Focus

An earthquake's **focus** is the exact point inside the Earth where the rocks first break and slip. These can be from several kilometers below the surface to several hundred kilometers deep.

Fault

The **fault** of an earthquake is a crack within the Earth where movement takes place. The rocks will slip along this surface.

Hanging Wall

This is the side of the fault that rests on the fault. In the diagram it is the left side. It is called the hanging wall because miners who would excavate along the inside of the fault would hang their lanterns on the overhanging wall.

Foot Wall

The **foot wall** is the side of the fault that is under the hanging wall. As miners walked along the inside of the fault, the rock that they walked on was called the foot wall. It is the right side of the diagram.

Normal Fault

In a **normal fault**, the hanging wall moves down and the foot wall moves up. The diagram shows a normal fault.

Reverse Fault

A **reverse fault** happens when the hanging wall is thrust upwards while the footwall sinks down.

Epicenter

An earthquake's **epicenter** is the spot on the surface closest to the focus. It will normally be the place where the shock is felt the strongest and the damage is the greatest.

Seismic Waves

Seismic waves are the vibrations that travel through the Earth. There are surface waves that travel along the surface of the Earth, and there are body waves that travel through the body of the Earth. Body waves travel out in all directions as expanding spheres. They can be bent (refracted) and/or reflected when they reach structures within the Earth.

When an earthquake occurs, the vibrations are called seismic waves. There are different types of seismic waves that vibrate in different ways.

However, even though they act differently and travel at different speeds away from the source of the earthquake, they all are made at the same time: the instant of the earthquake.

P-Waves

The **P-wave** is sometimes called the "primary wave" because it arrives at distant seismic stations first. Technically speaking, the P-wave is a compressional wave, which means that the rock material expands and contracts as the wave passes through. Here are the major distinctions of the P-wave:

- Phastest wave—so it arrives at seismic stations phirst.
- Push-pull wave—rock vibrates forward and backward in the same direction that the wave travels (propagates). It vibrates parallel to propagation.
- Passes through solids and liquids (magma of the interior).

To help remember the properties of the P-wave, notice that most of the descriptions of the P-waves start with the letter P.

S-Waves

- Secondary wave—arrives at a seismic station second.
- Slow wave—not as fast as the P-wave.
- Shake wave—vibrates side to side. In other words, if the wave travels from left to right, the vibration will be up and down—similar to ocean waves.
- Solids wave—only travels through solids.

To help remember the properties of the S-wave, notice that most of the descriptions of the S-waves start with the letter S.

Is the S-Wave the Same as an Aftershock?

No. P-waves and S-waves are made at the same time: The moment of the earthquake. They simply arrive at distant seismic stations at different times. However, at the epicenter of the earthquake, you would experience both waves at the same time.

An aftershock is a completely new earthquake. These often come in groups because the movement from the original earthquake will get the two sides of the fault moving.

The P-waves and the S-waves are the most important for studying Earth Science but there are other kinds of seismic waves. Specifically, they are called Love waves and Raleigh waves, named after the scientists who did extensive research on the waves. These two types of waves travel even slower than the S-wave. The main difference between these two waves and the P- and S-waves is that Love and Raleigh waves are called "surface waves." These waves travel along the surface of the Earth and can actually be seen sometimes as actual waves in the ground. The P-waves and S-waves are called "body waves" because they travel in all directions—most of which are down through the body of the Earth.

Lesson 4-5 Review

1. The focus of an earthquake can be defined as
   a) the point on the Earth's surface directly above the origin of the earthquake.
   b) any seismic station that detects and earthquake.
   c) a measure of the seismic wave strength.
   d) the point inside the Earth where the actual crustal movement takes place.
2. The differences between P-waves and S-waves are that
   a) P-waves are slower and travel through solids and liquids.
   b) P-waves are faster and travel through solids only.
   c) S-waves are faster and travel through solids and liquids.
   d) S-waves are slower and travel through solids only.
3. The earthquake will be felt strongest
   a) at the Moho.
   b) at the epicenter.
   c) where the P-wave and S-wave are furthest apart.
   d) in an airplane.
4. The hanging wall on a fault is
   a) whichever side goes up.
   b) the side of the fault that does the pushing.
   c) the side that you step on.
   d) the side that rests on top of the fault.
5. The earthquake wave with the highest speed is the
   a) S-wave. b) P-wave. c) seismic wave. d) Raleigh wave.

6. P-waves and S-waves are called body waves. The other type of seismic waves are
   a) sound waves.
   b) surface waves.
   c) compressional waves.
   d) longitudinal waves.

Lesson 4–6: Finding the Epicenter of an Earthquake

Earthquake waves (seismic waves) are detected and recorded by a device called a seismograph. The seismograph draws a picture of the vibrations that looks similar to a heartbeat called a seismogram. The seismograph will normally pick up some "background" vibrations from distant earthquakes, rock slides, trucks driving nearby, even waves crashing on the beach! The first indication that an earthquake has happened in a distant location is the arrival of the P-wave. Some time later, a second pattern is picked up, which will be the S-wave. The further away the seismic station is away from the earthquake focus, the more spread out the two waves will be. This information is used to find the exact location of the earthquake's focus.

Think of a race between two cars. One is a fast sports car and the other is a slow family car. If the race is a short distance the sports car will win, but not by much. If the race goes for a long distance, the sports car will win by a large margin. The same is true for the seismic waves: the further the waves travel, the more they get spread out.

Time Math

In order to do the calculations that help find the distance to the epicenter and the time of the earthquake, you'll need to do math with time. It is difficult to do time math in a calculator, and by hand does not work exactly the same way as regular math. Regular math is what they call "base 10," which means that whenever you count past 9 you must move over one place to the tens column. Time is base 60. You can count 59 seconds and then you go to the minutes column.

For example, if you want to solve 82 minus 17 in regular base ten numbers you would "borrow" a ten and start by taking 7 away from 12.

$$\begin{array}{r} 82 \\ \underline{-17} \end{array} \quad \text{is the same as} \quad \begin{array}{r} ^{7}\not 8\,^{1}2 \\ \underline{-17} \\ 65 \end{array} \quad \text{(70 plus 12)}$$

Time math works almost the same way, except instead of taking over ten from the neighboring column you'll take one minute and convert it into 60 seconds.

3:13:25 (3 hours, 13 minutes, and 25 seconds)

–1:09:37

turns into:

3:12:85 (3 hrs, 12 min, 85 secs is the same as 3:13:25)

–1:09:37

2:03:48

How to Use the P-Wave and S-Wave Travel Time Chart

The P-line shows how much time it takes a P-wave to travel a certain distance. So if you need to know how much time it takes the P-wave to travel 2,000 km, it is just over 4 minutes (about 4:05). The S-wave works the same way: For 2,000 km it takes 7:20.

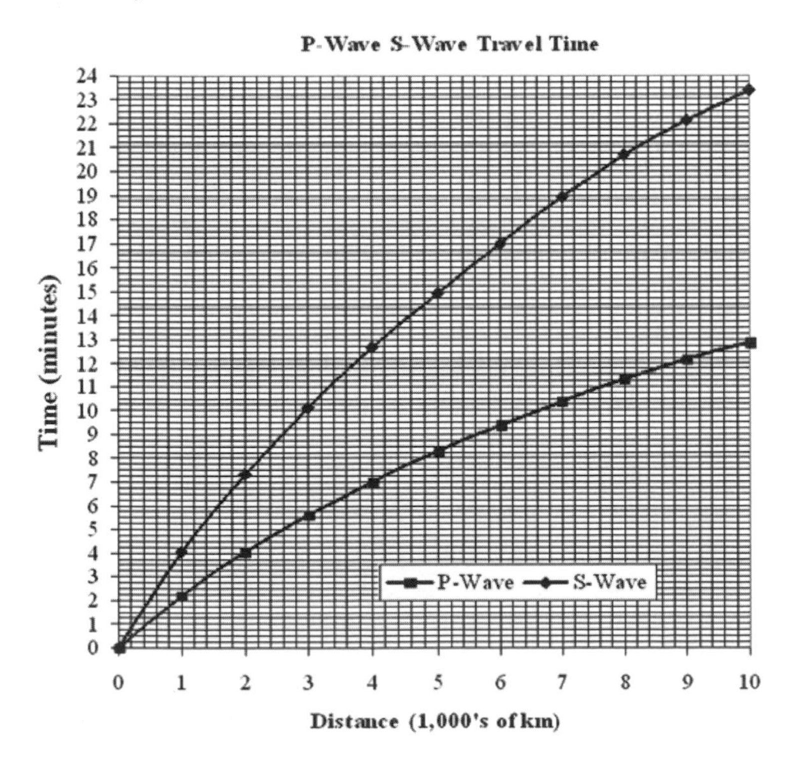

Figure. 4.11. P-Wave S-Wave Chart

To Find the Distance to the Epicenter

You are in charge of watching the seismic station tonight when the seismograph detects an earthquake. The earthquake didn't happen where you are—you can't even feel it. As a result, you don't know what distance or in what direction the earthquake occurred. The P-wave and S-wave are separated by 4:05 (4 minutes, 5 seconds). You need to find a spot on the graph where the P-line and the S-line are separated by 4:05. Line up a scrap piece of paper along the left edge of the chart. Put a small tick mark on your scrap paper at zero, and a small tick mark at 4:05. Now, slide the scrap paper up along the chart until the two tick marks just touch the P and S lines. *Be sure that your scrap paper is perfectly straight up and down* (use the lines on the grid as a guide). Now that you have found the right spot on the graph, drop a line straight down to the bottom of the graph to read the distance: 2,600 km.

To Find the Time That the Earthquake Occurred

When a seismograph detects an earthquake that happened at some distance, (2,600 km for example) you know that the earthquake happened some time in the past and it took time for the waves to reach your station. But how long ago? All you need to do is answer the question, "How long does it take a P-wave to travel 2,600 km?" Find 2,600 km on the bottom of the chart. Go straight up until you reach the P-line and read the time from the left of the chart: 5:00 (5 minutes). Now compare times. If you detected the earthquake at 3:17:00 and it took 5:00 then the earthquake happened 5 minutes before 3:17:00, or 3:12:00.

To Find the Epicenter on a Map

If you only have data from one seismic station, you know how far away the earthquake was, but not in what direction.

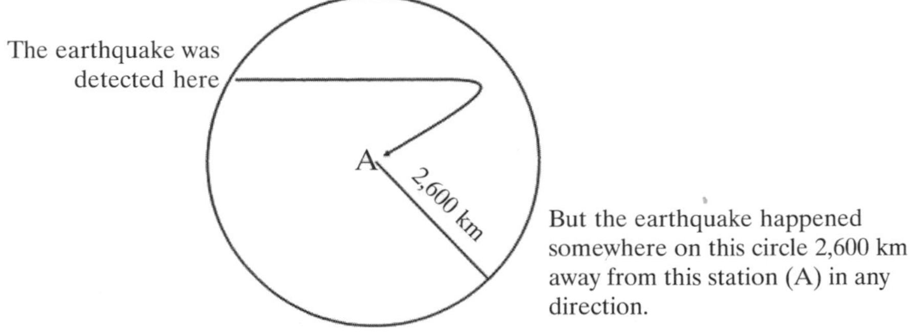

Figure 4.12. One station

If you have two seismic stations, their circles will agree on two possible locations for the earthquake epicenter.

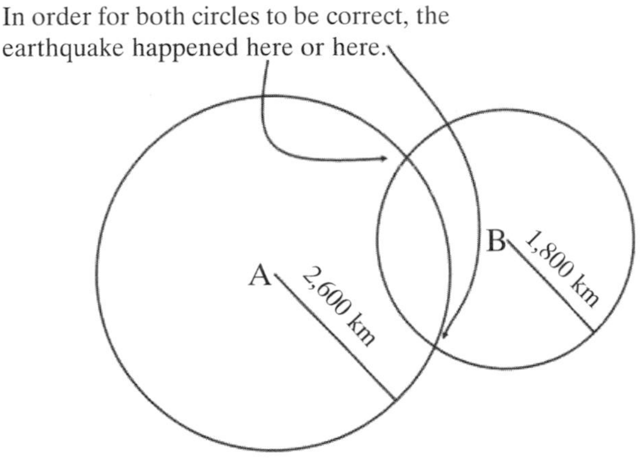

Figure 4.13. Two stations

You need the information from at least three seismic stations to find where the earthquake happened. The earthquake happened where all three circles intersect. Note: If you do this activity in class, it is normal to finish with a small triangular space for the epicenter instead of a clean intersection.

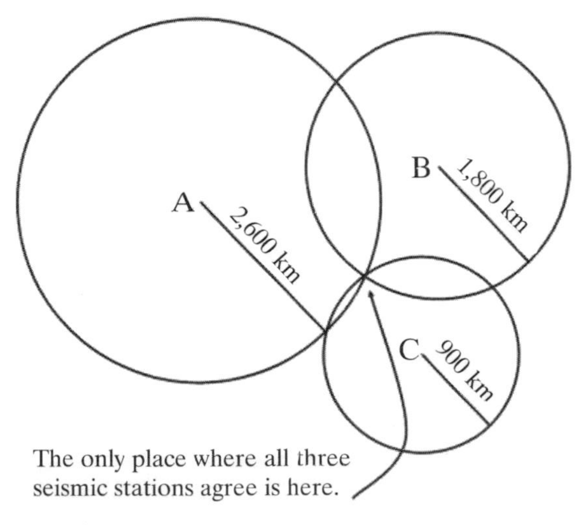

Figure 4.14. Three stations

To Sum Up the P-wave and S-Wave Chart

Distance to Epicenter

- Calculate the time difference between the arrival of the P-wave and the S-wave.
- Use the left side of the chart to put tick marks on scrap paper separated by the difference in arrival times.
- Slide your scrap paper up to the spot on the graph where the two lines are separated by that time (be sure that your paper is straight up-and-down).
- Once you find the spot on the graph, go straight down to read the distance.

Time of Earthquake

- Once you know the distance to the earthquake, consult the chart to see how long it takes a P-wave or S-wave to travel that distance.
- Subtract the travel time (the smaller number) from the arrival time.

Lesson 4–6 Review

Complete the charts below using the P-Wave and S-Wave Travel Time Chart on page 115.

Exercise A

seismograph station	Arrival (clock time)		Difference in arrival time (min. & sec.)	Distance to epicenter (km)	P-wave travel time (min. & sec.)	Time of origin (hr, min, & sec)
	P-wave	S-wave				
A	5:33:40	5:36:40				
B	5:40:25	5:48:45				
C	5:37:30	5:43:30				

Exercise B

seismograph station	Arrival (clock time)		Difference in arrival time (min. & sec.)	Distance to epicenter (km)	P-wave travel time (min. & sec.)	Time of origin (hr, min, & sec)
	P-wave	S-wave				
1	10:23:35	10:30:35				
2	10:23:20	10:30:00				
3	10:18:15	10:20:55				

Multiple Choice

1. What is the distance to the epicenter of an earthquake if the travel time for the P-wave is 4 minutes?
   a) 1.0×10^3 km
   b) 2.0×10^3 km
   c) 2.6×10^3 km
   d) 12.8×10^3 km

2. A seismograph receives a travel time difference of 7:00 (7 minutes) between P-waves and S-waves of an earthquake. Approximately how far is the seismic station from the epicenter of the earthquake?
   a) 1.9×10^3 km
   b) 4.0×10^3 km
   c) 2.9×10^3 km
   d) 5.4×10^3 km

3. a) What is the difference in arrival times of P-waves and S-waves at a seismic station 6,200 km away from the epicenter of an earthquake?
   b) If the earthquake occurs at 9:20:00 a.m., what time will a seismic station 2,600 km away pick up the P-wave?
   c) What time will it pick up the S-wave?

Lesson 4–7: Analysis of Seismic Waves

The intensity or strength of an earthquake is measured in two main ways. The first way, the Richter Scale, is the most commonly used for most people. It is the scale you will hear on the news whenever a major earthquake has happened.

The **Richter Scale** measures the amount of energy that is released by the earthquake, but it doesn't tell how much damage has happened. For example, two identical earthquakes happen: one in Manhattan and one in Antarctica. Even though both earthquakes are identical, the one in Manhattan will be a much more serious event than the Antarctic one.

The Richter Scale strength, or magnitude, is the number that gets reported for each earthquake. Each number of magnitude is 10 times stronger than the number below it. For example, a Richter magnitude 4 is 10 times stronger than a 3, and it is 100 times stronger than a 2 (10 × 10).

Because each number is a multiple of the one before, there is no real upper limit to the Richter Scale. But for all practical purposes, the Earth can only focus so much energy in one spot at one time. The strongest earthquake on record is the 1964 Chile earthquake, which was a 9.4.

On the other end of the scale, earthquakes with a magnitude of a 1 or 2 are constantly happening all over the world. They can be caused by trucks driving by, waves hitting the shore, and people falling down stairs as well as the regular way the earthquakes happen. These quakes are just too weak for us to normally feel them.

The other way that earthquakes are measured is with the Mercalli Scale (sometimes referred to as the "Modified Mercalli Scale"). The **Mercalli Scale** measures earthquake strength by classifying observations that people make during the earthquake.

This scale ranges from I to XII, with I being the weakest. The scale is based on common earthquake occurrences such as "noticeable by people," "damage to buildings," "chimneys collapse," and "fissures open in the ground."

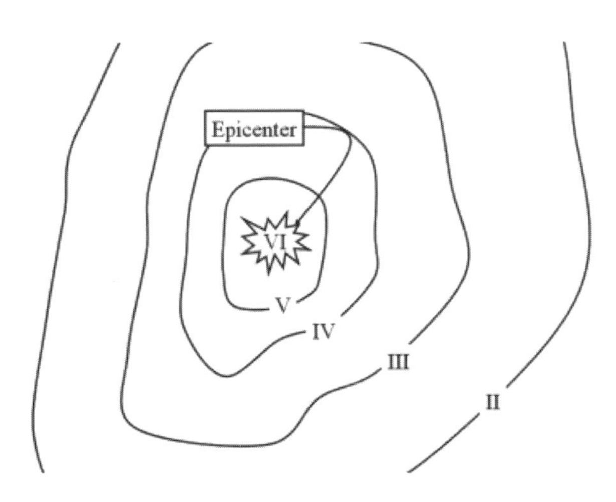

Figure 4.15 Mercalli Earthquake

Because of the way that the scale is designed, each earthquake will have several rankings. In Figure 4.15, notice that the lines of equal intensity are not perfectly round. This is normal and is caused mainly by the types of bedrocks in the area and the construction of the buildings.

Earth's Interior Structure

Despite what you may have seen in the movies, nobody has ever penetrated more than a few miles into the Earth to study the mantle or the core. We know about these deep places by studying seismic waves.

One critical piece of information comes from the difference in behavior between the P-wave and the S-wave. P-waves pass through liquids and S-waves do not. When a strong earthquake happens on the opposite side of the Earth, P-waves are detected, but S-waves are not. This tells us that there is a zone of liquid in between the two locations that the P-wave can travel through but the S-wave cannot. This liquid zone is the outer core.

Another interesting property of seismic waves is that they can be reflected off of sudden changes in properties such as the place where two layers meet. By measuring reflections, we can measure the depth down to the different layers. It works very much like a sonogram that a doctor might use to "look" at an unborn baby.

Finally, the speeds of seismic waves are well known to seismologists. The speed can change depending on the density of the material it is traveling through. The higher the density, the faster the speed of the wave. As a result, seismologists can calculate the density of each layer that the seismic waves pass through.

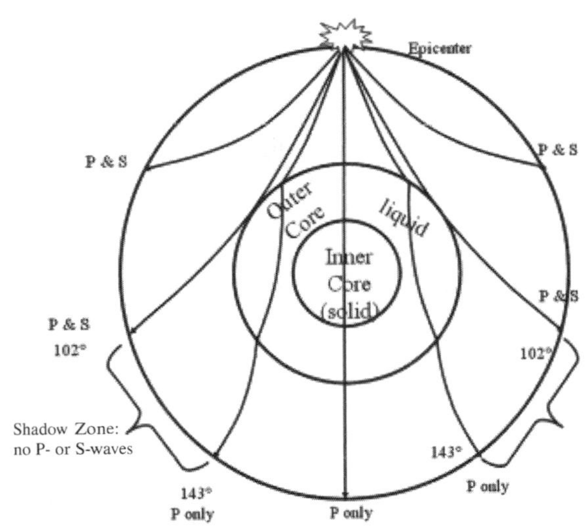

Figure 4.16 Shadow Zone

In Figure 4.16 you can see that the seismic waves get bent, or refracted, as they pass into deeper parts of the Earth. This is caused by one side of the wave being a little deeper than the other and, as a result, moving faster. It is just like trying to run when one foot goes a little faster than the other. You start running in circles.

When the wave hits the outer core, there is a sudden change in the wave's speed and there is a sharp turn. This diverts the wave into a new direction. Because of the deflection, there are no waves at all (P or S) detected between 102° and 143°around the Earth from the earthquake. This area, with no detectable seismic waves from a distant earthquake, is called the **shadow zone**.

Beyond the shadow zone—past 143°—only P-waves will be detected. This is because of the liquid outer core. As mentioned earlier, P-waves can travel through both solid and liquid but S-waves can only pass through solid.

The Tsunami

A tsunami is otherwise known as a tidal wave. Although it is commonly referred to as a tidal wave, a tsunami has nothing to do with tides. A **tsunami** is an ocean wave that is started by an earthquake on the sea floor. When an earthquake happens on the ocean floor and one side of the fault it shoved up while the other side drops, it will displace the water above. Such a disturbance will set up a "bouncing" of the water above the epicenter, creating waves.

Tsunamis are not significant while still out in the middle of the ocean. Far from shore, a tsunami may be two or three feet high. As it gets closer to shore it grows in height—sometimes to enormous heights. A tsunami can travel across an entire ocean and cause devastation thousands of miles away from the epicenter.

Lesson 4–7 Review

1. Besides the danger from the actual earthquake, an additional hazard caused by an earthquake that people in coastal communities have to deal with is

   a) a tsunami. b) a Raleigh wave. c) volcanic eruptions.

2. How much of a change is it to go from one number on the Richter Scale to the next?

   a) 1X c) 10X
   b) 2X d) It goes up by the square.

3. The strongest that an earthquakes can be on the Mercalli Scale is

   a) 10. b) XII. c) A. d) 1.

4. Which scale rates an earthquake based on the destruction it did?

   a) H-R Scale b) Fujita c) Richter d) Mercalli

5. The main reason why we know the outer core is made of a liquid is because
   a) seismic waves bounce off of the upper boundary of the outer core.
   b) lava that comes from a volcano is liquid.
   c) S-waves do not reach the other side of the Earth.
   d) P-waves get amplified in the outer core.

Chapter 4 Exam

Complete the charts using the P-Wave and S-Wave Travel Time Chart on page 115.

Exercise A

seismograph station	Arrival (clock time)		Difference in arrival time (min. & sec.)	Distance to epicenter (km)	P-wave travel time (min. & sec.)
	P-wave	S-wave			
1	2:05:25	2:08:00			
2	2:13:15	2:22:20			
3	2:08:15	2:13:05			

Exercise B

seismograph station	Arrival (clock time)		Difference in arrival time (min. & sec.)	Distance to epicenter (km)	P-wave travel time (min. & sec.)
	P-wave	S-wave			
A	1:19:20	1:22:40			
B	1:21:15	1:25:30			
C	1:20:35	1:25:05			

Exercise C

1. List four pieces of evidence that Pangaea existed.
2. If a P-wave arrived at a seismic station at 2:45:20 and the S-wave arrived at 2:51:00, how far away was the epicenter of the earthquake?

3. What is formed at a divergent plate boundary?
   a) Pangaea
   b) subduction
   c) high, folded mountains
   d) new crust
4. Which layer of the Earth is liquid?
   a) crust b) mantle c) outer core d) inner core
5. Most of the Earth's seismic activity is concentrated
   a) in the middle of the Atlantic Ocean.
   b) in the middle of the Pacific Ocean.
   c) around the Pacific Ocean.
   d) in Antarctica.
6. Which layer of the Earth is the densest?
   a) crust b) mantle c) outer core d) inner core
7. Hawaii is a large volcano with gently sloping sides made by lava flows. It can be classified as which type of volcano?
   a) cinder cone
   b) composite cone
   c) shield volcano
   d) explosive volcano
8. What is the minimum number of seismic stations needed to find the epicenter of an earthquake?
   a) 1 b) 2 c) 3 d) 4
9. The spot inside the Earth where an earthquake begins is called the
   a) hanging wall. b) fault. c) epicenter. d) focus.
10. What will happen in a collision between an ocean plate and a continental plate?
    a) The continental plate will override the ocean plate.
    b) The oceanic plate will override the continental plate.
    c) Both plates will subduct.
    d) Both plates will rise up.
11. Hawaii does is not on the edge of a plate and yet it is very active. Hawaii is an example of a
    a) hot spot.
    b) divergent plate boundary.
    c) lithospheric plate.
    d) rift valley.

Answer Key

Answers Explained Lesson 4–1

1. **Inner core, outer core, mantle, crust**

   Alternate answer: **Inner core, outer core, mantle, asthenosphere, oceanic crust, continental crust**

2. **A** is incorrect because the oceanic crust is solid rock but it has liquid water sitting on top of it.

   B is **correct** because the outer core is liquid iron and nickel.

   C is incorrect because the inner core is solid iron and nickel.

   D is incorrect because the continental crust is solid rock.

3. **B** is **correct** because the chemical symbols on the Periodic Table of Elements for Iron and Nickel are "Fe" and "Ni."

   A, C, and **D** are incorrect.

4. **A** is **correct** because the Moho discontinuity is one of the places where properties change drastically. (The other places are where two layers meet such as the outer core and mantle.)

   B, C, and **D** are incorrect because properties within the mantle and outer and inner cores change very gradually as you go deeper.

5. **A** is **correct** because the continental crust is mostly granite.

   B is incorrect because the oceanic crust is mostly basalt.

   C and **D** are incorrect because magma and lava are molten rock. Once they solidify into solid rock, they are no longer called magma and lava.

6. **A** is incorrect because the continental crust is mostly granite.

   B is **correct** because the oceanic crust is mostly basalt.

   C and **D** are incorrect because magma and lava are molten rock. Once they solidify into solid rock, they are no longer called magma and lava.

7. **C** is **correct** because continental crust is thinner and lighter than oceanic crust which is why the continents stick up high enough to be above water.

   A, B, and **D** are incorrect.

Answers Explained Lesson 4–2

1. **A** is incorrect because geologic activity can happen in populated as well as remote areas.

   B is incorrect because, although some activity can happen in the centers on continents, it is concentrated on the edges.

   C is **correct** because geologic activity happens where two plates rub against each other or are getting torn apart.

D is incorrect because Antarctica has some geologic activity, but not the most in the world.

2. **A** is incorrect because all three activities happen near each other and along the edges of the plates.

   B is incorrect because these events still happen every day.

   C is incorrect because these events have little effect on the weather.

   D is **correct** because all three events happen near each other.

3. **A** is incorrect because there are many active volcanoes in populated areas.

   B is **correct** because most volcanoes are found at the edges of the plates.

   C is incorrect. In fact, the majority of active volcanoes in the world are under the ocean.

   D is incorrect because Hawaii is not on the edge of a plate.

4. **A** is **correct** because it gets its name from the fire from frequent volcanic eruptions.

   B is incorrect because the Ring of Doom was Frodo's ring of power.

   C is incorrect because a hot spot is located in the center of a plate. The Pacific Rim follows the edge of a plate.

   D is incorrect because Dante's Peak is the name of one volcano.

Answers Explained Lesson 4–3

1. **C** is **correct** because oceanic plates are denser than continental. As a result, the oceanic plates will sink while the continental will rise up.

   A, **B,** and **D** are incorrect.

2. **A** is incorrect because continental crust is made of granite, which is lighter than the basalt that makes up oceanic crust.

   B is **correct** because oceanic plates are denser than continental because they are made of basalt, which is denser than the continental granite.

   C is incorrect because continental crust has a lower density than oceanic crust.

   D is incorrect because transform is not a type of plate, it is a type of plate boundary.

3. **A** is incorrect because crust is often destroyed at a convergent plate boundary.

   B is **correct**. When the plates spread apart, lava wells up to form new crust.

   C is incorrect because crust is neither created nor destroyed at a transform plate boundary.

   D is incorrect.

4. **A** is **correct**. When oceanic and continental plates converge, the oceanic plate will subduct. As it sinks, it will pull down the ocean floor, creating a trench.

B is incorrect because new crust will be formed at a divergent plate boundary. (The crack that the lava comes out of is not deep and it is filled with lava.)

C is incorrect because transform plate boundaries do not have trenches.

D is incorrect.

5. **A** is **correct**. The ocean plate will sink because it is denser.

   B is incorrect because the continental plate is not denser.

   C is incorrect because the mountains stick up so high because of their low density.

   D is incorrect because when continental and oceanic plates converge, the denser, oceanic plate will always sink.

6. **A** is incorrect because two plates that are moving apart are divergent.

   B is **correct** because transform plates slide sideways.

   C is incorrect because two plates that are moving toward each other are convergent.

   D is incorrect because a boundary between plates where new crust is created is a divergent plate boundary.

7. **A** is incorrect because isostatic rebound happens when the land rises after a heavy load such as a glacier is removed.

   B is incorrect because seafloor spreading happens at a divergent plate boundary. The question is referring to a convergent plate boundary.

   C is **correct** because *subduction* is the term for a plate that gets forced under another.

   D is incorrect because *rifting* is another word for splitting that happens at a divergent plate boundary.

8. **A** is incorrect because the Mid-Atlantic Ridge is very active.

   B is incorrect because the Mid-Atlantic Ridge is a divergent plate boundary.

   C is incorrect because this happens at a convergent plate boundary. The Mid-Atlantic Ridge is a divergent plate boundary.

   D is **correct** because the ridge is formed from the volcanoes and new crust welling up from the divergent crack.

Answers Explained Lesson 4–4

1. **A** is incorrect because Gondwanaland was a smaller "super continent."

   B and **C** are incorrect because Pangaea was the name of the supercontinent.

   D is **correct**.

2. **A** is **correct** because this is not evidence that Pangaea existed. Although this may be proof that the plates are colliding against each other creating friction, it is not evidence.

   B is incorrect because this is evidence. Magnetic stripes on the ocean floor show that the floor of the ocean is splitting along the Mid-Atlantic Ridge.

   C is incorrect because this *is* evidence. Similar fossils on opposite shore are very good evidence that the continents were together.

   D is incorrect because this is evidence. The "puzzle fit" of the continents is a compelling piece of evidence that the continents fit together at one time.

3. **A** is incorrect because there is no evidence that the tilt of the world has significantly changed in the past.

   B is incorrect because when continents collide it happens slowly by our standards (an inch or two a year). In addition, if coal was to somehow bounce during the collision, it would only be found right on the shore.

   C is **correct** because Antarctica was once closer to the equator and has since migrated over the South Pole—carrying its coal with it.

   D is incorrect. Even though heat and pressure can form coal, Antarctica would need the raw ingredients to start with: abundant plant material.

4. **A** is **correct** because India was closer to the South Pole and actually attached to Antarctica at one time.

   B is incorrect because there is no evidence that the world was cold enough so that continental glaciers could exist near the equator. Alpine (or valley) glaciers may have existed, but not continental.

   C is incorrect because Antarctica is cold because it is at the South Pole, not because it is Antarctica. If Antarctica was at a warmer latitude it would be just as warm as any other continent.

   D is incorrect because magnetic reversals do not affect the weather or climate.

5. **A** is incorrect. Even though both oceans have magnetic stripe patterns, both of their patterns are different from each other.

   B is incorrect. In the Atlantic Ocean, the stripes parallel the shores of the Atlantic—which is roughly up and down. Lines of latitude run side to side.

   C is incorrect because the magnetic patterns are not connected with the fossils.

   D is **correct** because the stripes are made at the Mid-Atlantic Ridge and are split apart which results in a mirror image.

6. **A, B,** and **C** are incorrect.

   D is **correct**. As you get closer to the center of the ocean, you get closer to where the sea floor is being formed. It gets younger. Once you pass the Mid-Atlantic Ridge the sea floor gets older.

Answers Explained Lesson 4–5

1. **D** is **correct**.

   A is incorrect because the point on Earth's surface directly above the origin of the earthquake is the epicenter.

   B and **C** are incorrect.

2. **D** is **correct**.

   A, B and **C** are incorrect.

3. **A** is incorrect because the Moho is very deep inside the Earth.

   B is **correct** because the epicenter is the spot on the surface that is closest to the source of the earthquake.

   C and **D** are incorrect.

4. **A** is incorrect. In a normal fault, the hanging wall goes down and in a reverse fault, the hanging wall goes up.

   B is incorrect because both side push against each other.

   C is incorrect because the side that you step on is the toe wall.

   D is **correct.**

5. **A** is incorrect because the S-wave is the slower wave.

   B is **correct** because the P-wave is faster.

   C and **D** are incorrect.

6. **A** is incorrect.

   B is **correct**.

   C is incorrect because P-waves are a type of compressional wave.

   D is incorrect because S-waves are a type of longitudinal wave.

Answers Explained Lesson 4–6

Exercise A

seismograph station	Arrival (clock time)		Difference in arrival time (min. & sec.)	Distance to epicenter (km)	P-wave travel time (min. & sec.)	Time of origin (hr, min, & sec)
	P-wave	S-wave				
A	5:33:40	5:36:40	**3:00**	**1,800**	**3:40**	**5:30:00**
B	5:40:25	5:48:45	**8:20**	**7,000**	**10:25**	**5:30:00**
C	5:37:30	5:43:30	**6:00**	**4,400**	**7:30**	**5:30:00**

Exercise B

seismograph station	Arrival (clock time)		Difference in arrival time (min. & sec.)	Distance to epicenter (km)	P-wave travel time (min. & sec.)	Time of origin (hr, min, & sec)
	P-wave	S-wave				
1	10:23:35	10:30:35	**7:00**	**5,200**	**8:35**	**10:15:00**
2	10:23:20	10:30:00	**6:40**	**5,000**	**8:20**	**10:15:00**
3	10:18:15	10:20:55	**2:40**	**1,500**	**3:15**	**10:15:00**

Multiple Choice

1. **A** is incorrect. If you got this answer you looked at the S-line instead of the P- line.

   B is **correct**. Go up the time scale on the left side of the chart to the 4 minute mark. Then slide across until you reach the P-line and slide straight down to read the distance.

   C is incorrect. If you got this answer you found the spot on the chart where the P-line and the S-line are separated by four minutes.

   D is incorrect.

2. **A** is incorrect. If you got this answer, you measured how far the S-wave traveled in 7 minutes.

   B is incorrect. If you got this answer, you measured how far the P-wave traveled in 7 minutes.

   C is incorrect.

   D is **correct**. You start by marking off two tick marks separated by 7 minutes by using the left side of the chart. Then slide your scrap paper with the tick marks up the chart and find a spot where the P- and S-lines are separated by the distance between the two tick marks.

3. a) **7:50**. Go to 6,200 km on the bottom of the chart. Line up a piece of scrap paper with the 6,200 km line. Put a tick mark on the scrap paper where it crosses the P-line and the S-line. Drag the scrap paper with the tick marks over to the side of the chart and measure the time.

   b) **9:25:00**. To travel 2,600 km, it takes the P-wave 5:00 (5 minutes) to reach the seismic station. 9:20:00 + 5:00 = 9:25:00

   c) **9:29:00**. It takes the S-wave 9:00 minutes to travel 2,600 km. 9:20:00 + 9:00 = 9:29:00

Answers Explained Lesson 4–7

1. **A** is **correct**.

   B is incorrect because a Raleigh wave is another seismic wave that will affect all areas struck by an earthquake, not just a coastal community.

   C is incorrect.

2. **C** is **correct**. Each level of strength is 10 times stronger than the one below.

3. **B** is **correct**.

4. **A** is incorrect because the H-R Scale is used in astronomy.

   B is incorrect because the Fujita scale rates tornadoes.

   C is incorrect because the Richter scale measures the amount of energy released during the earthquake.

   D is **correct.**

5. **A** is incorrect because seismic waves bounce off of all of the layers within the Earth, not just the liquid or solid ones.

   B is incorrect because the lava that comes from a volcano does not come from as deep as the outer core.

   C is **correct** because S-waves cannot pass through liquid.

   D is incorrect.

Answers Explained Chapter 4 Exam

Exercise A

seismograph station	Arrival (clock time)		Difference in arrival time (min. & sec.)	Distance to epicenter (km)	P-wave travel time (min. & sec.)
	P-wave	S-wave			
1	2:05:25	2:08:00	**2:35**	**1,500**	**3:15**
2	2:13:15	2:22:20	**9:05**	**7,800**	**11:10**
3	2:08:15	2:13:05	**4:50**	**2,600**	**5:00**

Exercise B

seismograph station	Arrival (clock time)		Difference in arrival time (min. & sec.)	Distance to epicenter (km)	P-wave travel time (min. & sec.)
	P-wave	S-wave			
A	1:19:20	1:22:40	**3:20**	**2,000**	**4:05**
B	1:21:15	1:25:30	**4:15**	**2,800**	**5:20**
C	1:20:35	1:25:05	**4:30**	**3,000**	**5:40**

Exercise C

1. **Puzzle fit of the continents.**

   Fossils of the same species on opposite shores.

   Mineral chains that continue across the oceans.

   Mountain chains that continue across the oceans.

   Glacial evidence.

   Magnetic stripes on the ocean floor.

   Continental crust is older than oceanic crust.

2. **The earthquake was 4,000 km away.**

   The difference between the P-wave and the S-wave was 5:40 seconds.

3. **A** is incorrect because Pangaea was an ancient supercontinent.

   B is incorrect because subduction happens where plates are colliding.

   C is incorrect because folded mountains are formed where plates are colliding.

   D is **correct** because the rift between the plates allows magma to well up and form new crust.

4. **C** is **correct**.

5. **A** is incorrect. There is some activity at the Mid-Atlantic Ridge but not as much as in the Pacific.

   B is incorrect because the middle of the Pacific Ocean is in the middle of a plate, which is not very active (except for Hawaii's gentle lava flows).

   C is **correct**. The edge of the Pacific Ocean is called the Ring of Fire.

   D is incorrect.

6. **A** is incorrect because the crust floats on top of all of the other layers inside the Earth.

B and **C** are incorrect because the mantle and outer core have densities in-between the crust and the inner core.

D is **correct** because the heaviest (densest) material sank to the center of the Earth.

7. **A** is incorrect because cinder cones are made of cinders—small volcanic rocks.

   B is incorrect because composite cones are made from cinders and lava. Hawaii is made from lava only.

   C is **correct**.

   D is incorrect because Hawaii is not an explosive volcano.

8. **A** is incorrect because one station will give you a distance but not a direction.

   B is incorrect because two stations will give you two possible locations for the epicenter.

   C is **correct** because three circles will cross at only one point: the location of the epicenter.

   D is incorrect because three is the minimum.

9. **A** is incorrect because the hanging wall is one side of the fault.

   B is incorrect because a fault is the crack along which movement takes place during an earthquake.

   C is incorrect because the epicenter is the spot on the Earth's surface closest to the focus.

   D is **correct**.

10. **A** is **correct** because the continental crust is less dense than the oceanic plate.

    B is incorrect because oceanic plates are too dense to override the continental plate.

    C and **D** are incorrect because the denser plate will sink and the lighter plate will rise.

11. **A** is **correct**.

    B, **C**, and **D** are incorrect.

Weathering and Erosion

Lesson 5–1: Physical Weathering

Weathering is the process that breaks rocks down when they are exposed to weather: rain, temperatures, and chemicals. Weathering will reduce rock material down into sediments based on properties of the rock, such as mineral composition, shape, and its environment. There are two different types of weathering: physical and chemical.

Physical weathering is the actual attacking of the rock material with physical force. Some examples can be hitting, scratching, and rubbing. **Chemical weathering** is, as it sounds, a chemical reaction between the minerals in the rock and other substances that it comes in contact with.

Types of Physical Weathering

Frost action, sometimes called frost wedging, is caused by the expansion of water within tiny cracks during the freezing process. Freezing water can exert tremendous pressure—enough to break solid rock. Water seeps into small cracks in rocks. When the water freezes it expands creating great pressure. The crack widens and allows water to seep deeper into the rock. Frost action works best in areas where the temperature fluctuates above and below freezing often. However, it does not work very well at the poles, as many people might think, because the water freezes once and stays frozen for many years. Frost action is the cause of most potholes in roadways.

Root action is similar to frost action—at least in the way that it works. When plants grow, tiny roots will search out cracks in the rock. As the plant grows, the root will expand, which will put pressure on the rock and crack it. Pressure of a growing root can be strong enough to crack through very large rocks. To help this process along, many plant roots secrete chemicals that weaken the minerals in the rock.

Exfoliation (ex = "outside"; foliation = "flaking") causes rocks to lose their outer surfaces by flaking off in layers. This is caused by extreme temperature changes. Because the outside of a rock is subjected to bigger swings in temperature than the inside, the outside will expand and contract the most, which will cause it to crack. Exfoliated rocks will often have the appearance of a partially peeled onion.

Pressure unloading is similar to exfoliation because it has to do with the rock changing its size. Many rocks are formed deep underground under extreme pressure. When that pressure is released, the rock will expand and, because it is not flexible, the expansion will cause it to crack. The outside of the rock will fall off in tiny pieces.

Abrasion is the rubbing of one rock against another. There are many ways in nature where two pieces of rock can rub against each other. For example: during a landslide; by the force of glaciers; by running water; or by being picked up by wind.

Lesson 5–1 Review

1. Name three types of physical weathering:

   a) ______________________________

   b) ______________________________

   c) ______________________________

2. True or False. Expansion and contraction of water is sufficient to cause rocks to crack.

3. Any process that forces a rock to crack or break into pieces is called
   a) dissolution.
   b) abrasion.
   c) physical weathering.
   d) chemical weathering.

4. The effects of repeated freezing and thawing of water is called
   a) root action.
   b) frost action.
   c) exfoliation.
   d) equilibrium.

5. Most materials will ____________ when freezing; an unusual feature of water is that it will ____________ when freezing.
   a) expand, contract
   b) contract, expand
   c) expand, expand
   d) contract, contract

6. The effects of frost action are most noticeable in places where the temperature usually
   a) is colder in the shade than in the Sun.
   b) stays above the freezing point of water.
   c) stays below the freezing point of water.
   d) fluctuates above and below freezing.

7. Roots growing into rocks can
   a) exert enough pressure to crack the rocks.
   b) hold the rocks together, preventing weathering.
   c) protect the rocks from extremes of weather.
   d) pressurize the rock, making it harder to weather.

Lesson 5–2: Chemical Weathering

Chemical weathering is caused by a substance coming into contact with the minerals of a rock and causing a chemical reaction. Chemical weathering can be very selective by attacking one mineral in a rock and leaving the surrounding minerals intact.

Oxidation is, as the name implies, a chemical reaction involving oxygen. One of the most common forms of oxidation is rust. Oxygen in the atmosphere or water combines with iron in the rocks creating rusty red iron oxide. Usually when you see a red-colored rock or soil, the color will come from iron oxide.

Water dissolution is the process of soluble minerals dissolving into water. Water has the nickname "the universal solvent" because it can dissolve just about anything else to some degree. However, some materials dissolve very easily in water, while others dissolve very slowly. Two common minerals that dissolve easily in water are calcite and halite (rock salt).

Acid comes in many forms, both natural and artificial. One natural source of acid is carbonic acid. Carbonic acid is formed when CO_2 (carbon dioxide) in the atmosphere combines with rain water to form a weak acid.

Because carbon dioxide is a natural gas in the atmosphere, pure rain water is normally slightly acidic. Carbonic acid is also known as seltzer—the same seltzer that is in soda.

The term **acid rain** usually refers to elevated levels of acid caused by human activity. Two examples of acid rain are sulfuric acid and nitric acid, both of which have their natural sources, but the problem comes from the elevated levels. Sulfuric acid rain is caused by sulfur dioxide (SO_2) in the atmosphere mixing with rain, which makes a strong acid. The acid rain will fall downwind of industry that burns large amounts of coal. A natural source of sulfuric acid is the rain that falls from volcanic clouds. The effects of sulfuric acid can be seen when it reacts with copper to make a grayish green coating such as on the skin of the Statue of Liberty. Nitric acid, another form of acid rain, is a bi-product of burning fossil fuels such as coal and gasoline.

Lesson 5–2 Review

1. Rust is a common example of
   a) physical weathering.
   b) chemical weathering.
   c) frost action.
   d) abrasion.

2. Chemical weathering is
   a) the result of scratching two substances together.
   b) the result of severely changing the material's temperature.
   c) the chemical composition of the rock material.
   d) the result of a reaction between two substances.

3. Because water can dissolve just about anything, it has the nickname
   a) the universal solvent.
   b) the water hammer.
   c) water acid.
   d) acid rain.

4. Carbonic acid is commonly known as
   a) acid rain.
   b) seltzer.
   c) copper sulfate.
   d) effervescence.

5. The sulfuric acid in acid rain
   a) comes only from human pollution.
   b) comes only from natural sources.
   c) comes from both natural sources and human pollution.
   d) is only found in industrial areas.

Lesson 5-3: Weathering Rates

Weathering rates will be influenced by surface area exposed, mineral composition of the rock, and climate conditions.

Surface Area Exposed

Because the attack of a rock by weather can only happen on the surface, the exposed area is a critical factor to how fast the rock will break down. For example, if you had a cube of rock material sitting on your desk and poured acid on it, the acid would have to eat away the outside first before it could attack the inside. But, if you took that same rock and cut it into several pieces, the acid would now have an opportunity to work on both the outside and the inside surfaces at the same time.

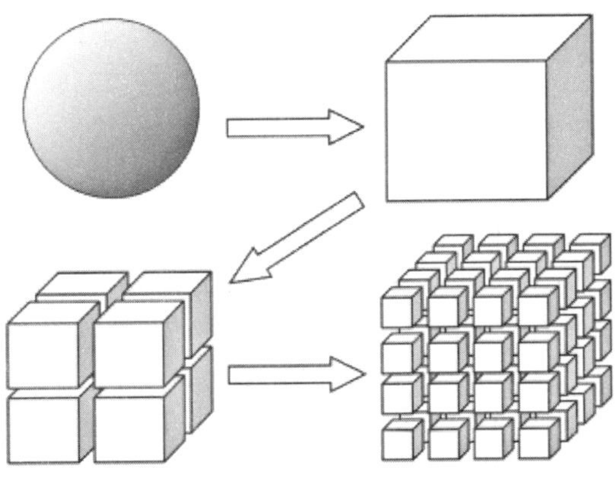

Figure 5.1. Surface area

Mineral Composition

Mineral composition is an obvious factor in why some rocks break down quickly. Some minerals are soft and will weather rapidly, while other minerals are resistant. Besides the hardness of the mineral, other properties such as reaction to chemicals, can also influence the weathering rates. For example, feldspar is a strong, tough mineral (hardness of 6), but it can be broken down fairly quickly (in rock terms) by carbonic acid. On the other hand, quartz which has about the same hardness (7) as feldspar, is resistant to most chemicals as well as being very hard. This is why most bottles used to store chemicals are made out of quartz glass.

Climate Conditions

The climate to which a rock is exposed can have profound consequences on the lifetime of a rock. If the rock is exposed to extremes in temperature, it may be subjected to physical weathering. If it is exposed to high temperatures and moist conditions, chemical weathering may break down the rock faster. If the rock is exposed to windy conditions or flowing water, it may be warn down by abrasion.

Climatic Conditions

- Cold and/or dry climates favor physical weathering.
- Warm and wet climates favor chemical weathering.
- Frost action works best in areas where the temperature fluctuates wildly.

Lesson 5–3 Review

1. True or False. The deeper a rock is, the more it will be affected by weathering.
2. True or False. The weathering rate of rocks is directly proportional to their size.
3. If a block of rock and a pile of sediment made from the same exact rock-type, with an identical weight as the block, were both left exposed to the weather, what would be the result after several years?
   a) The block will break down faster.
   b) The pile of sediment will lose more mass than the block.
   c) The two samples will weather at exactly the same rate because they are made of the same material.
   d) The sediments will reform into a solid rock.
4. Office buildings will often have polished marble on the building faces. Even though marble is relatively soft and reactive with acid rain, the marble weathers slowly because
   a) the smooth surface has a lower surface area than a rough surface.
   b) the polished surface reflects harmful sunlight.
   c) only buildings in clean areas with stable climates use marble.
   d) only the marble that is exposed higher than neighboring buildings will be weathered.

Lesson 5–4: Soil

Soil is a mixture of sediments, water, air, and organic material. The sediments come from the broken-down remains of rocks. The organic material in soil comes from decayed and decaying plants and animals as well as animal waste such as manure.

Soil is the product of weathering. If you start with a mountain of solid rock, eventually the mountain will be weathered into sediments and mix with water, air, and organic material. The soil of each area is unique to that area because it is a combination of all the factors that have produced it, such as the original rock and the climate.

The Evolution of Soil

Soil begins as an exposure of solid rock (1). The surface of the rock gets attacked by the weather and broken into smaller pieces (2). Over many years, you wind up with heavily weathered sediments at the surface. As you dig deeper, you find less and less weathering and the materials become coarser (3).

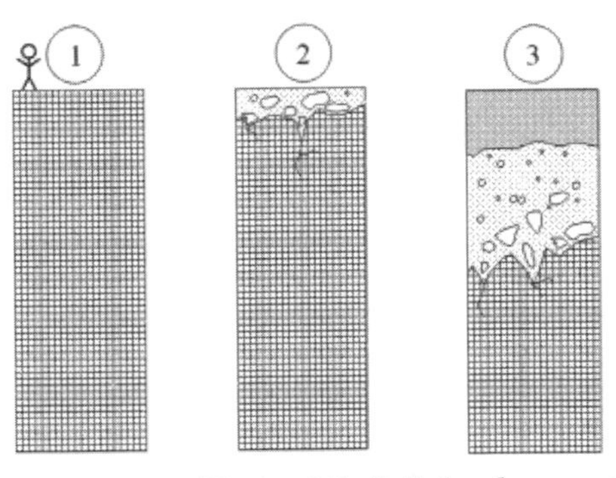

Figure 5.2. Soil development

When the soil becomes mature, it will start to collect organic material in the form of decayed plant and animal material, animal waste, bacteria, and other organisms such as worms. This rich and dark mixture of material, called **humus**, makes the top layer fertile and able to support plant life.

A column of soil is called a "soil profile."

As you start digging down from the top of the soil profile, the first layer, which is the top layer, is called the **topsoil**. Topsoil is usually dark and fertile. The next layer lower in the profile is the sub-soil. The sediments in this layer are essentially the same as the top soil but there is no organic matter. As a result, this layer is lighter in color. Below the sub-soil is the location where the bedrock is being broken up. This is the layer of partly weathered bedrock. Here you will find larger sediments and rocks. Just below the partly weathered bedrock is the original bedrock that formed the soil above.

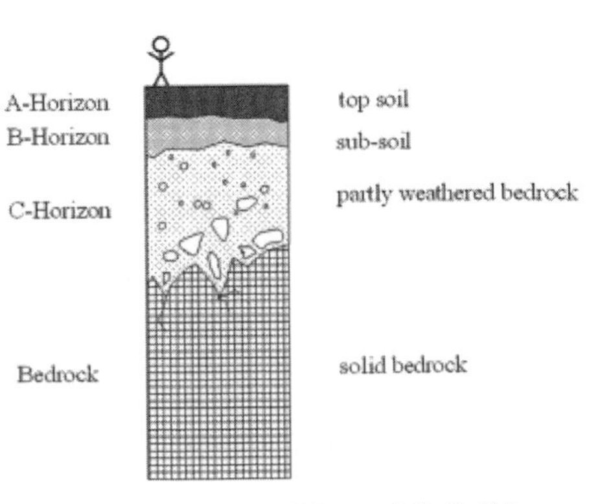

Figure 5.3. Soil layers

Rather than refer to the layers by their long names, soil scientists refer to the layers with letters. The topsoil is the A, sub-soil is the B, and the partly weathered bedrock is the C. Because each layer is horizontal, the layers are called the A-horizon, B-horizon, and the C-horizon.

Residual and Transported Soils

The previous story of the evolution of soil is the simplest version of how bedrock can be broken down into soil. This is the case with residual soil. **Residual soil** is soil that has been made from the bedrock directly below it. Even though this represents the simplest formation of soil, it is actually less common. The reason for this is simple: It takes so long for soil to form, and there are so many opportunities for the sediments to be moved by water or wind, that it often gets carried off.

Residual soil gets its name because it lives, or *resides*, with its parent rock.

Transported soil is soil that has been made from transported sediments and is the more common type of soil. Basically, this soil has been formed by the weathering down of the high places (mountains and hills) and the sediments making their way down into the low places (valleys and plains). The sediments can get pushed along by running water, glaciers, or just by gravity alone. The sediments of transported soil do not match the minerals of the bedrock below.

Lesson 5–4 Review

1. Soils developed from material different from the bedrock are known as _____________soils; those developing on top of their parent bedrock are _____________ soils.
2. True or False. Transported soils are more common than residual soils.
3. The layer of soil with the most organic material is the
   a) bedrock. b) C-horizon. c) B-horizon. d) A-horizon.
4. The sediments that make up soil often come from
   a) valleys. b) mountains. c) plains. d) the ocean.

5. The main difference between sub-soil and topsoil is
   a) topsoil is residual.
   b) sub-soil has more organic material.
   c) topsoil has more organic material.
   d) topsoil is transported.

6. The organic material in soil is also known as
   a) sub-soil. b) humus. c) bedrock. d) the profile.

Lesson 5–5: Erosion

Erosion is the process of moving sediment from one location to another. Erosion should not be confused with weathering (the breaking down of rock) ,but the two processes almost always go together.

The processes of weathering and erosion on the surface of the Earth have one overriding goal: to make the surface flat and level. The only reason why nature hasn't reached this goal yet is because the forces inside the Earth are constantly reworking the crust. It's this battle between the forces that build up the Earth and those that tear it down that creates the diversity in our landscapes.

When the internal forces of our planet push the land upwards, creating mountains, the rocks at the mountain peaks are supplied with energy—potential energy. **Potential Energy** is the form of energy an object gets because of its position—in this case, a high elevation. A rock at the top of a mountain may not be doing anything now, but it has the potential to do something in the future: fall. The reason why a rock will fall is because, at some time in the past, the energy required for it to fall was stored within the rock when it was moved to its current height. In short: A rock that has a long way to fall has a lot of potential to fall (and break and crash and break other rocks) whereas a rock sitting on low-lying level ground does not have much chance of falling very far.

When an object falls, it gets lower to the ground and the potential energy decreases. It decreases because, from moment to moment, it has less distance to fall. But the potential energy doesn't disappear; it gets converted into kinetic energy. **Kinetic energy** is the energy of motion. When an object (or the agent pushing it around) has a lot of kinetic energy, erosion will occur. Eventually, all of an object's potential and kinetic energy will be lost as friction and the sediments will stop moving and get deposited.

The most important force of erosion is gravity. Anything that moves sediments around is somehow caused by gravity. Water and glaciers move downhill. Wind and ocean currents are moved by density differences which are driven by gravity.

An **agent of erosion** is the thing that actually does the pushing. Some agents of erosion can be running water, glaciers, wind, and people. Gravity by itself can also be an agent of erosion, such as in a rock fall or landslide. Of all the agents of erosion, running water is the most important.

A stream will transport material in a few ways: in suspension, rolling and bouncing, dragging, and in solution. When sediment is carried in **suspension**, it is carried within the flow of water. This luxury is usually reserved for the smallest and lightest particles. As the particles get larger or denser, the water may not have the pushing power to completely lift the particle, in which case it will bounce along the bottom. If it is round enough, the particle will roll. The largest particles that the stream can move will slide along the bottom. Finally, a stream can transport a fair amount of material in solution. This will be minerals such as halite (salt) and calcite that are soluble in water.

The amount of material transported or eroded by a stream, called its load, depends on several factors. However, the two most important factors are velocity and volume of water flow. **Velocity** of a stream is simply the speed of the water. It makes sense that the faster the water flows, the more material it can push along and the larger material can be pushed along.

The velocity of a stream depends on the amount of water flowing through the stream channel and the slope of the channel. **Discharge** is the term for the amount of water that flows through a stream. It represents the amount of water that flows past a point in one minute. It could be measured in gallons per minute (gal/min) or cubic meters per minute (m^3/min). Discharge is essentially the "faucet setting" and depends on the amount of rainfall or groundwater available.

Think of discharge as the stream being a large pipe and the water in the stream is discharging out of the end of the pipe.

The slope of the stream is the inclination or gradient of the stream bed. Of course, water will slip down a very steep slope much faster than a gentle slope. Even though the inclination will make the water flow faster, it has no effect on the amount of water that flows. The faucet still remains the same.

Stream Flow and Particle Size

The size of sediments that can be transported by a stream depends mainly on the velocity of the stream. The swifter the water, the larger the particles will be that get moved.

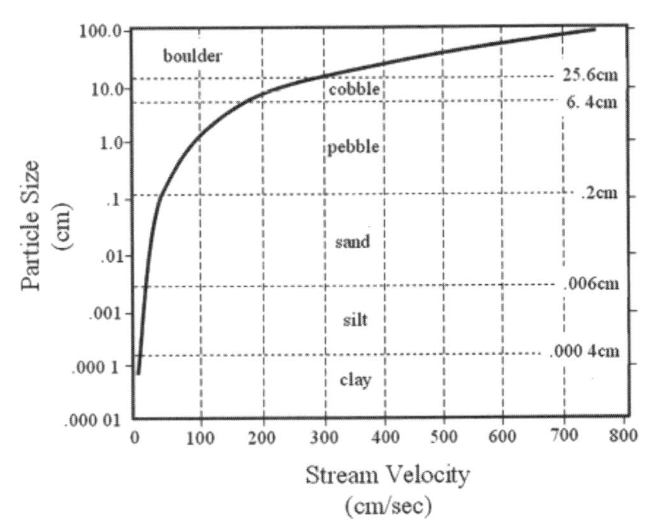

Figure 5.4. Stream velocity

The previous chart shows the relationship between the velocity of the water and the size of the sediment that it can move. This is the speed that the water needs to flow to transport the particle. In order to start the motion, the water's speed will need to be even faster to overcome friction with the stream bed.

The stream velocity chart assumes that there is enough water to move the sediments. No matter how fast the water is flowing, if there is simply not enough water, it will not be able to move as much. Likewise, properties of the particles such as density and shape may adjust the relationships of the chart.

Lesson 5–5 Review

1. Stream discharge is also known as
   a) velocity. b) volume. c) capacity. d) slope.
2. A steeper gradient will ____________ the velocity of a stream.
   a) increase b) decrease c) disrupt d) not affect

3. Erosion is
   a) the same thing as weathering.
   b) the breaking down of rock material.
   c) the transportation of sediments.
   d) the most important agent of weathering.
4. A steeper gradient will ____________ the volume of the stream.
   a) increase b) decrease c) disrupt d) not affect
5. The most important agent of erosion on the surface of the Earth is
   a) gravity.
   b) running water.
   c) ice (glaciers).
   d) wind.
6. Objects that are resting at high elevations have a lot of
   a) potential energy.
   b) kinetic energy.
   c) dynamic equilibrium.
   d) surface area.
7. A heavy rainfall will increase a stream's
   a) discharge. b) slope. c) potential energy.
8. Kinetic energy increases as
   a) minerals react with acid rain.
   b) a rock gets weathered into smaller pieces.
   c) a sediment gets higher in the mountains.
   d) the object picks up speed.
9. How fast must a stream flow in order to transport a particle 1.0 cm in size?
   a) 50 cm/sec b) 90 cm/sec c) 125 cm/sec d) 300 cm/sec
10. A stream moving with a velocity of 300 cm/sec will transport which set of sediments?
    a) silt and clay
    b) clay, silt, sand, and most pebbles
    c) clay, silt, sand, pebbles, and most cobbles
    d) clay, silt, sand, pebbles, cobbles, and some boulders up to 75 cm wide

Lesson 5–6: Stream Erosional-Depositional Systems

Deposition is the process of materials getting deposited after erosion. The patterns of deposition reveal the environment of the area when the sediments were deposited. Whenever erosion happens, deposition will eventually happen somewhere else.

When a stream (or any agent of erosion) loses kinetic energy, it slows down and loses its ability to transport sediment. As it slows, the stream begins to drop the sediments that are hardest to transport. The first sediments to come out of transport are the roundest, largest, and densest particles. The last sediments deposited will be the flattest, smallest, and lightest particles.

Horizontal Sorting

When a stream slows down gradually, it will gradually drop its sediment load roughly in size order, with the largest falling first. As the stream continues onward, it drops the next largest and then the smaller and smaller particles. In the quietest water, the smallest sediments are deposited and then even the dissolved minerals deposit as a precipitate. This pattern of dropping sediments one at a time in size order is called **horizontal sorting**. You will get horizontal sorting wherever the velocity of a stream gradually slows down. This can happen when a stream widens or when it enters a quiet body of water such as a lake or an ocean.

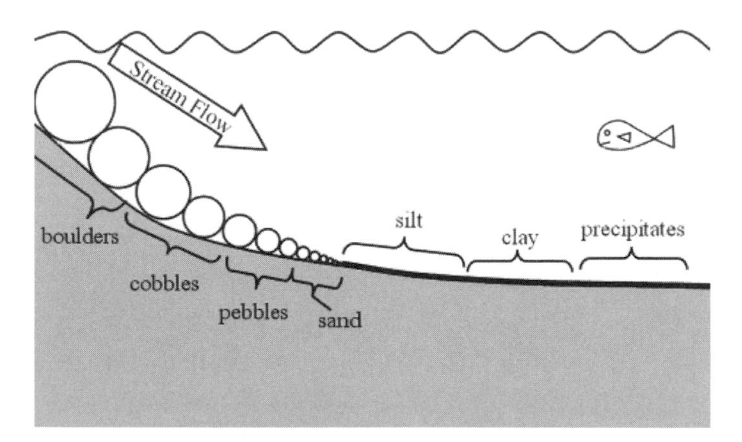

Figure 5.5. Horizontal sorting

Vertical Sorting

Vertical sorting is another way that water can drop materials in size order. In **vertical sorting**, the largest, heaviest particles are at the bottom of a stack because they were dropped first, and the smaller particles are on top. Vertical sorting will happen whenever there is a sudden stop in water velocity. This will typically happen when there is a storm or flood that suddenly increases velocity, picks up a large mixture of sediments, and then dies off. All of the sediment picked up by the flood waters is suddenly dropped in size order. If sorted layers stack up from cyclic depositional events such as spring thaw, you can get a pattern called **graded bedding**.

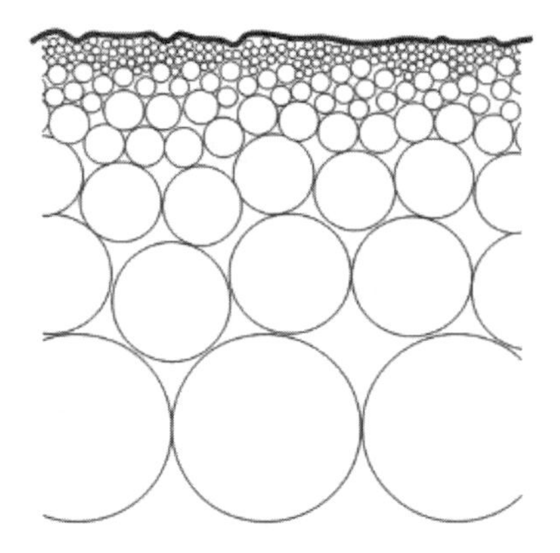

Figure 5.6. Vertical Sorting

Stream Shape and Erosion and Deposition

The shape of a stream will determine where erosion and deposition will occur. In a straight stream channel, the fastest water will be above the center of the water flow. Along the sides, there will be friction that will slow the stream down. As a result, the channel will be deeper in the center, where the fastest water is located, and shallower at the edges, where the flow is minimal. A stream valley is called a **v-shaped valley** because the sides collapse into the river as erosion widens the banks. This produces a v-shaped slope on the two sides of the stream.

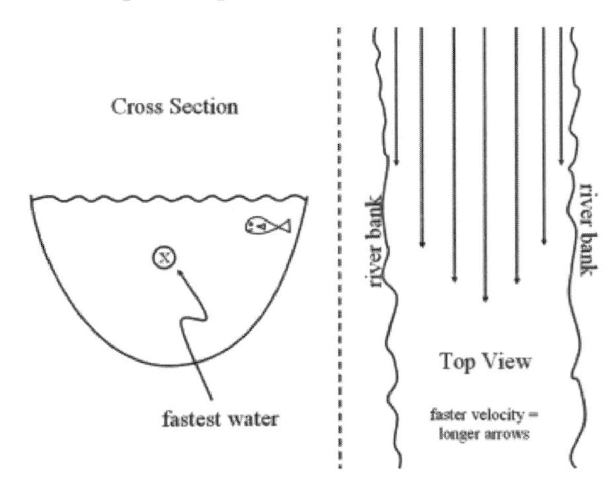

Figure 5.7. Stream speed straight

Where the stream curves, the fastest water gets thrown to the outside of the turn and the slowest stays to the inside. Think of it as a car race on a slick and icy track. The cars that are fastest get thrown to the outside and scrape against the outer wall. Therefore, the outside of the curve will be where the most erosion happens. For this reason, the outside curve is sometimes called the "cut-bank." The inside of the turn, where the velocity is slowest, is where deposition will happen.

If you ever go panning for gold, the best place to prospect is the inside curves of a stream because gold is dense and gets deposited first wherever the water slows.

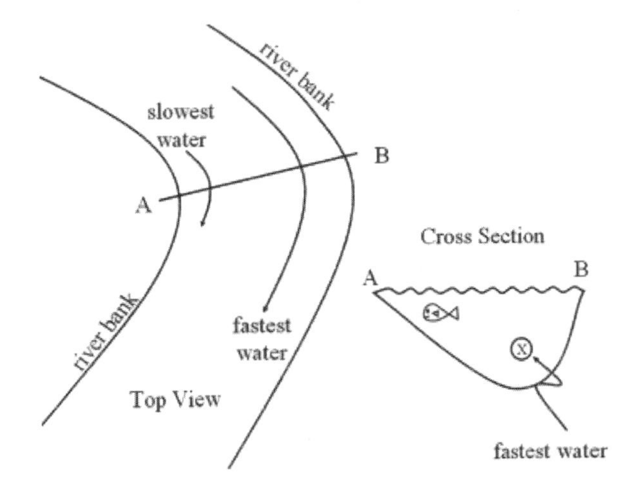

Figure 5.8. Stream speed curved

As a stream develops, any small turns start to get exaggerated because of the erosion and deposition happening on the sides of the stream. The outside turns get rounder and wider and develop into features called meanders. A **meander** is a wide turn in the stream bed. Meanders can bend so much that one side of the turn touches the other side. When this happens, the stream will jump the bank and follow the easiest path, stranding the meander, leaving a cut-off **oxbow lake** to the side.

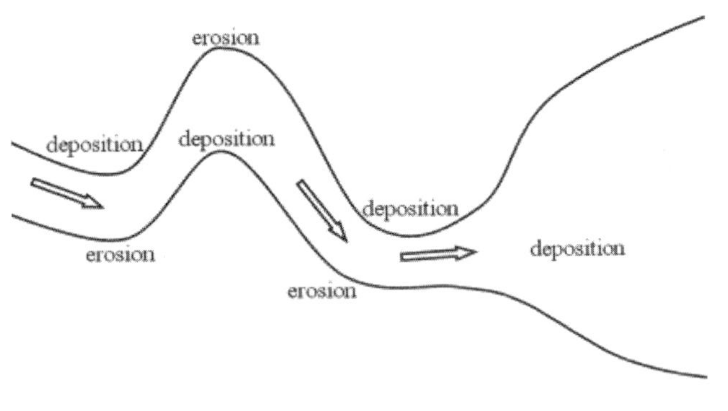

Figure 5.9. Meanders

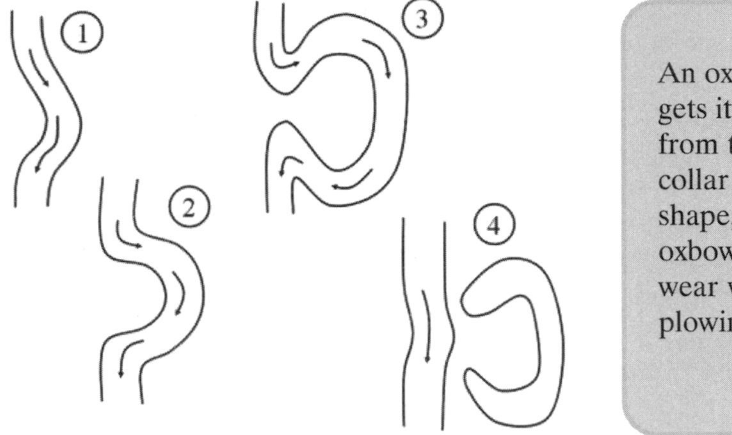

> An oxbow lake gets its name from the large collar of similar shape, called an oxbow, that oxen wear when plowing fields.

Figure 5.10. Oxbow

The Life of a Stream

Rivers go through stages in the evolution of their development which are similar to the stages people go through: youth, maturity, and old age. Each stage is marked by the typical characteristics of the stream.

Youth

A youthful stream is characterized by an energetic flow. Swift water caused by a steep gradient results in a lot of erosion—which is the dominant process. A youthful stream will have a narrow v-shaped valley from the collapse of the stream walls caused by rapid down-cutting.

Maturity

When a youthful stream gets well established, the side walls of the stream will continue to collapse and widen the v-shape of the stream. Lateral ("to the side") erosion begins and the stream valley widens. When the valley is wide enough, meanders will develop. At this point, the valley will be wider than the actual water channel, which leaves room for a flood plain. During flood stages, the river will overflow its banks and cover a wide but shallow shoulder called the **flood plain**. A mature stream has a gentler gradient and slower water.

Old Age

This stage is characterized by numerous and wide meanders and oxbow lakes. The land is almost flat and the water flows slowly. The river snakes and wanders around because the very wide flood plain has no hills to steer the river. Erosion and deposition balance each other in dynamic equilibrium. Whatever sediment is eroded from the outside of the channel, the same amount is deposited on the opposite side.

Keep in mind that one stream often has many different ages, depending on which section of the stream you are observing. Near the headwaters high in the mountains, the stream will be youthful and energetic. In the lowlands, the same stream may be old and slow. The sections also do not need to be in age order. In some cases, an old slow river can go over a cliff into a waterfall and become youthful again.

Lesson 5–6 Review

1. After a sudden flood, sediments will be deposited in
   a) horizontal sorting.
   b) vertical sorting.
   c) the outside of meanders.
   d) sub-soil.
2. You can find the fastest water and the most erosion
   a) at the edges of a straight section of the stream.
   b) at the river delta.
   c) on the outside of a turn.
   d) at the bottom of the deepest part of the stream.
3. Over time, a meander of a stream will tend to
   a) straighten out.
   b) get more exaggerated.
   c) erode downward to make a deeper channel.
   d) fill in with cobbles.
4. High in the mountains, a river will usually be
   a) youthful.
   b) mature.
   c) old-aged.
   d) a combination of youthful and old-aged.
5. Youthful streams typically have
   a) v-shaped valleys.
   b) meandering valleys.
   c) u-shaped valleys.
   d) wide valleys.

6. When a stream has a balance between erosion and deposition, it has reached
   a) the source of the river.
   b) the mouth of the river.
   c) static equilibrium.
   d) dynamic equilibrium.

7. The outside of a curve will have the ____________ flowing water.
   a) fastest
   b) slowest
   c) shallowest
   d) clearest

8. A gradual slowing of stream velocity will deposit sediments in a pattern called
   a) vertical sorting.
   b) horizontal sorting.
   c) meandering.
   d) cross-beds.

9. Flattest materials get deposited
   a) first.
   b) fastest.
   c) slowly.
   d) in fast water.

Lesson 5–7: Other Erosional-Depositional Systems

Agents of Erosion

The characteristics of deposited particles give information about the erosion and deposition environments. This is one of the clues to look for in deposits or sedimentary rocks that tell us the environment of deposition.

Running Water

Sediments transported by, and deposited in, running water will often be rounded and smooth. The typical "river stone" is a good example. The sediments will usually be sorted. Vertical sorting tells of a sudden stop to the water such as a storm. Horizontal sorting tells of a gradual slowing such as where a river enters a larger body of water.

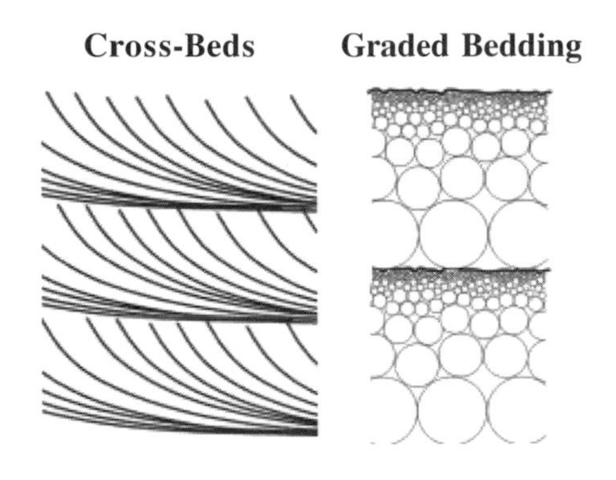

Figure 5.11

Multiple layers often develop from seasonal events, such as a spring flood or summer drought. This will add a layer of vertical sorting on top of another layer of vertical sorting in a pattern called **graded bedding**. In a place where the river channel meanders back and forth across itself there can be cross-beds.

Cross-beds are slanted layers deposited in river deltas or meandering stream beds. Each layer in a cross-bed shows horizontal sorting. A **delta** is a cone-shaped deposit at the mouth of a stream. The word *delta* comes from the Greek letter delta, which has a triangular shape similar to the river deposit.

Wind

Wind works very similar to water except on a smaller scale. The sediments picked up by wind are fine, not usually larger than a grain of sand. Wind deposits are usually sorted; sometimes the fine sediments have cross-beds. Sediments eroded by wind will often have a frosted or pitted appearance from being sandblasted. Solid rock in sandblasted areas will show the erosion in the bottom few inches where the sand is most concentrated.

Ice (Glaciers)

The sediments transported by glaciers can vary greatly in size. Boulders the size of houses, called erratics, are commonly found after a period of glaciation. When a glacier picks up sediment, the particles get embedded into the ice and weather very little during transport. When the sediments are deposited as the ice melts, they look sharp and angular. When the ice dumps its sediment load, it all drops at once—unsorted. Unsorted glacial deposits are called **till**. Key clues that tell us of glacial deposition are the sharp, angular sediments that are unsorted and any size from very large to very small.

At the bottom of the glacier, where the dirty, sediment-laden ice scrapes along the surface, there can be considerable friction. Here you will see parallel scratches in the bedrock. The bedrock can get polished by the abrasive action of the glacier.

A **continental glacier** is also called an ice sheet. This type of glacier, as the name suggests, covers entire continents. Currently there are only two in the world: Antarctica and Greenland. There are numerous features associated with continental glaciers:

- **Outwash plain:** a very broad and flat area where glacial meltwaters deposited sediments—horizontally sorted.
- **Drumlin:** oval-shaped hills that are parallel to the direction of travel of the glacier. They will have a blunt end on the side where the glacier hit it and a long, gentle slope on the other side pointing the way that the glacier traveled.

- **Kettle lake:** a round, deep lake formed from a chunk of the glacier that got embedded into the outwash plain. As the ice-berg melted, the fine sediments within it lined the hole which allowed it to retain the water to form a lake.
- **Terminal moraine:** a line of glacial till that marks the furthest extent of the glacier.
- **Recessional moraine:** While the glacier melts back during warmer climates, it may pause and leave a new line of till. Recessional and terminal moraines are similar except the terminal marks the furthest extent.
- **Esker:** Sometimes called a "raised stream bed," it is a long, skinny hill. It was formed by a river of meltwater that flowed at the bottom of the glacier.
- The landscape will be littered with large transported boulders called **erratics**.

The other type of glacier is the **alpine** or **valley glacier**. These are glaciers that flow like slow moving, frozen rivers in mountain valleys. One of the main features associated with this type of glaciation is the **u-shaped valley**. Alpine glaciers occupy the mountain valleys in which liquid water used to flow before the cold climate. The heavy, powerful ice scrapes the sides and widens the v-shaped channel of the running water into a wider u-shape.

Gravity

By itself, gravity can act as an agent of erosion and deposition as well as the main force of erosion. Gravity will erode material by pulling it downhill without the help of anything else. This can be in the form of a landslide or a rock falling from a cliff. These sediments will be sharp-edged because they didn't weather down very much as they fell. Sediments will be unsorted and have no limit in size.

Gravity erosion is not always a sudden event. It can also work slowly. Over the course of many years, soil will slowly creep down hill because of the pull of gravity. Moving entire hillsides slowly over time is a type of mass wasting and is called **downhill creep**.

Clues in Deposited Sediments

- **Water**
  - Rounded polished sediments.
  - Horizontal sorting = gradual slowing.
  - Vertical sorting = sudden stop.
  - Cross-bedding = river delta or meandering stream.
- **Wind**
  - Frosted, pitted sediments.
  - Sediments are small.
  - Sediments are sorted.
- **Glaciers**
  - Unsorted and greatly varying in size.
  - Frcshly weathered (sharp and angular).
  - Parallel scratches on bedrock.
- **Gravity**
  - Sharp and angular sediments.
  - Unsorted, greatly varying in size .

Lesson 5–7 Review

1. Downhill creep is a type of mass wasting caused by
   a) running water.
   b) wind.
   c) glaciers.
   d) gravity.
2. The dominant agent of erosion on the Earth is/are
   a) running water.
   b) wind.
   c) glaciers.
   d) gravity.
3. Sediments transported by wind will have a "frosty" appearance because of ______________ in the surface.
   a) pits
   b) layers
   c) crystals
   d) minerals

4. A boulder moved by a glacier is called
   a) a delta. b) a moraine. c) a meander. d) an erratic.
5. A sudden rush of rocks and soils down a slope is called
   a) a landslide.
   b) downhill creep.
   c) meandering.
   d) dynamic equilibrium.
6. Sediments that have been transported by running water will typically be
   a) sharp and angular.
   b) small, frosted, and pitted.
   c) rounded and smooth.
   d) all sizes and can be very large.
7. The cone-shaped deposit at the end of a stream is a
   a) delta. b) gamma. c) meander. d) cut bank.
8. Graded bedding has several layers, each of which show
   a) parallel scratches. c) vertical sorting.
   b) bench marks. d) large crystals.

Chapter 5 Exam

Exercise A

Put a P next to the terms that describe *physical weathering* and a C next to examples of *chemical weathering*.

1. _____ Two rocks scratching at the bottom of a glacier.
2. _____ Acid rain dissolving the calcite in marble.
3. _____ H_2O expanding within cracks when it undergoes a phase change.
4. _____ Rust forming on exposed iron-rich rocks.
5. _____ Roots from a tree splitting rocks in half.
6. _____ The Statue of Liberty's skin turning green after years of exposure.

Exercise B

Classify the following particle sizes:

7. .07 cm ________________

8. .005 cm ________________

9. 20.5 cm ________________

10. 4.2 cm ________________

11. .0002 cm ________________

12. 39.5 cm ________________

Exercise C

Determine what particles would be carried by a stream moving at the following velocities:

13. 50 cm/sec ________________

14. 100 cm/sec ________________

15. 200 cm/sec ________________

16. 250 cm/sec ________________

17. 300 cm/sec ________________

Multiple Choice

18. Which of the following is NOT done by water when it freezes?
    a) It gets denser.
    b) It expands.
    c) It exerts pressure.
    d) It solidifies.

19. The natural acidity in rain comes from
    a) evaporated sea water.
    b) carbon dioxide (CO_2).
    c) pollution.
    d) minerals dissolved in rain water.

20. Chewed candy dissolves faster in your mouth faster than leaving the candy intact because
    a) unchewed candy has a more surface area.
    b) the act of chewing dissolves some of the candy.
    c) chewing the candy changes its resistance to scratching.
    d) the chewed candy has a larger surface area.

21. Warm and wet climates favor
    a) physical weathering.
    b) chemical weathering.
    c) frost action.
    d) wind abrasion.

22. The main difference between topsoil and sub-soil is that
    a) sub-soil is usually darker.
    b) sub-soil has smaller sediments.
    c) topsoil has humus.
    d) topsoil is made of mineral sediment.

23. One difference between transported soil and residual soil is that transported soil
    a) stays in the same location as its parent rock.
    b) is more common than residual soil.
    c) has more humus than residual soil.
    d) is found on river banks.

24. As a sediment makes its way to the bottom of a mountain it will
    a) gain potential energy.
    b) get larger.
    c) lose potential energy.
    d) recrystallize.

25. An earthquake changes the landscape and makes a river bed steeper by tilting it in place. One result will be
    a) decreased potential energy.
    b) increased volume.
    c) increased potential energy.
    d) increased stream velocity.

26. The natural process of rock transportation is called
    a) oxidation.
    b) graded bedding.
    c) weathering.
    d) erosion.

27. Pieces of rock fragments weathered off of rocks are called
    a) minerals. b) sediments. c) cross-beds. d) deposits.

Answer Key

Answers Explained Lesson 5–1

1. **frost action**
   root action
   exfoliation
   pressure unloading
   abrasion

2. **True**. The expansion of water is sufficient to easily break rock.
3. **A** is incorrect. Dissolution means to dissolve a rock. Similar to sucking on candy, the rock will get smaller but not normally get broken into pieces.

   B is incorrect. Abrasion is just one type of process. It is not "any" process.

   C is **correct**. Physical weathering will result in broken pieces of rocks.

   D is incorrect. Chemical weathering will not force a rock into broken pieces.
4. **A** is incorrect. Root action involves plants and not the freezing and thawing of water.

   B is **correct**. Frost action is the expansion and contraction of water when it freezes and thaws.

   C is incorrect. Exfoliation is a peeling of rocks caused by heating and cooling but does not involve water.

   D is incorrect. Equilibrium means "balance" and does not directly relate to repeated freezing and thawing.
5. **B** is **correct**. Water is unusual in the way that it expands and gets lighter when freezing; everything else contracts and gets denser. This is why ice floats on top of the liquid water.
6. **A** is incorrect.

   B is incorrect because water will not freeze if the temperature stays above the freezing point of water.

   C is incorrect. Even though freezing temperatures will make ice, frost actions needs many freeze-thaw cycles in order to work.

   D is **correct**. In order for frost action to work best, it needs to be repeated many times.
7. **A** is **correct**. Roots are strong enough to split rocks apart.

   B is incorrect. While holding the rocks together, roots will crush and split the rocks.

   C is incorrect. The roots themselves are not enough protection. A layer of soil will help protect rocks from extremes.

   D is incorrect. Roots cannot increase the resistance of rocks.

Answers Explained Lesson 5–2

1. **A** is incorrect because rusting is a chemical reaction.

   B is **correct**. Rusting is the chemical reaction of oxygen combining with iron.

   C is incorrect because frost action is the weathering caused by water expanding within cracks.

   D is incorrect because abrasion is the physical process of rubbing rocks together.

2. **A** is incorrect because scratching two substances together is an example of physical weathering.

   B is incorrect because severely changing the material's temperature may result in physical weathering. It will not alter the mineral's chemistry.

   C is incorrect because the chemical composition of the rock material is not a weathering process; it is just a description.

   D is **correct**. Chemical weathering is the result of a reaction between two substances.

3. **A** is **correct**.

   B is incorrect.

   C is incorrect.

   D is incorrect. Acid rain refers to a chemical within impure water.

4. **A** is incorrect because acid rain can be any form of acidic precipitation.

   B is **correct**.

   C is incorrect. Copper sulfate forms when sulfuric acid rain combines with copper.

   D is incorrect. Effervescence is the bubbling that happens when a strong acid attacks a substance.

5. **C** is **correct**. Sulfuric acid can come from pollution as well as volcanic activity.

Answers Explained Lesson 5–3

1. **False**. The deeper a rock is, the more it will be protected from the elements of weather by a covering of soil and other rock.

2. **False**. The weathering rate of rocks is inversely proportional to their size. The bigger the rock, the slower the weathering rate will be.

3. **A** is incorrect because the block has a smaller surface area than the pile of sediments (all of the sediments added together).

   B is **correct** because the pile of sediments has a larger surface area than the solid block.

   C is incorrect because whichever sample has the higher surface area exposed will break down faster—even if they are the same material.

   D is incorrect because sediments need to be deeply buried for a very long time in order to re-form into a sedimentary rock.

4. **A** is **correct**. A rough surface has little bumps on it that increases its surface area.

   B is incorrect. Even though the polished surface will reflect some light including harmful light, rocks are not broken down quickly by exposure to sunlight.

C is incorrect because buildings all over use marble facings and have the same results. Polished marble hold up longer than rough marble.

D is incorrect because neighboring buildings will offer very little protection.

Answers Explained Lesson 5–4

1. Soils developed from material different from the bedrock are known as **transported** soils; those developing on top of their parent bedrock are **residual** soils.

2. **True**. In the lifetime of a soil, chances are really good that something will come along and move it.

3. **A** is incorrect because the bedrock has no organic material in it. Bedrock is solid rock.

   B is incorrect because the C-horizon is partly weathered bedrock (pebbles and sand).

   C is incorrect because the B-horizon is sub-soil and sub-soil is the same as topsoil except that sub-soil has no organic matter in it.

   D is **correct** because the A-horizon is topsoil which contains all of the organic material in a soil.

4. **A** is incorrect because sediments only move down due to gravity.

   B is **correct**. Gravity pulls weathered materials downwards until they get to a relatively low spot.

   C is incorrect because plains are flat. Once sediments get to these areas, they will have a difficult time getting transported to any place that is lower.

   D is incorrect because the ocean is the ultimate final resting place for sediments. (If they get uplifted because of tectonic forces, they will either be solid rock by then or the sediments will slide down hill as the uplift is happening.)

5. **A** is incorrect because topsoil can be either residual or transported—just as sub-soil can be.

   B is incorrect because sub-soil has less organic material.

   C is **correct** because the topsoil is the layer that has organic material.

   D is incorrect because sub-soil can be either residual or transported—just as topsoil can be.

6. **A** is incorrect because sub-soil has no organic material.

   B is **correct**. Humus is the organic material that is found in topsoil.

   C is incorrect. Bedrock is the solid rock beneath soil.

   D is incorrect because a soil's profile is the column of layers as seen from the side.

Answers Explained Lesson 5–5

1. **A** is incorrect because velocity is the speed of the stream.

   B is **correct** because discharge is the amount of water, or volume, of a stream.

   C is incorrect because capacity is the amount of material that a stream can carry.

   D is incorrect because slope is the gradient of the stream bed.

2. **A** is **correct** because a steeper gradient will make the water slide downhill faster.

   B is incorrect because a steeper gradient makes water flow faster.

   C is incorrect because a steeper gradient will help a stream's flow, not disrupt it.

   D is incorrect because increasing the gradient will definitely affect the stream flow.

3. **A** is incorrect because weathering and erosion are two different processes (but they usually work together).

   B is incorrect because the breaking down of rock material is weathering.

   C is **correct**.

   D is incorrect because erosion is not an agent of weathering.

4. **A, B,** and **C** are incorrect because slope has no affect on the amount of water flowing.

   D is **correct** because changing the slope of a stream will not change how much water flows (the "faucet setting").

5. **A** is incorrect because gravity is a minor agent of erosion. However, it is the most important *force* of erosion.

   B is **correct** because running water is the dominant thing that pushes sediments around the surface.

   C is incorrect. Even though glaciers are very strong and have had a major impact on much of Earth's landscape, running water affects everywhere at some time.

   D is incorrect because wind does not have as much strength as water.

6. **A** is **correct** because potential energy has to do with an object's height.

   B is incorrect because objects that are resting have no energy of motion.

   C is incorrect because a resting object will have a static equilibrium.

   D is incorrect because surface area has nothing to do with elevation.

7. **A** is **correct** because discharge is the amount of water that flows through a stream.

   B is incorrect because slope will not be affected by rainfall.

C is incorrect because potential energy has to do with the elevation of the stream.

8. **A** is incorrect because reaction with acid rain has nothing to do with kinetic energy.

   B is incorrect because weathering into smaller pieces increases surface area.

   C is incorrect because it is potential energy that increases as a sediment gets higher in the mountains.

   D is **correct** because kinetic energy is energy of motion, which increases as the object picks up speed.

9. **B** is **correct**. Go up the left side of the stream velocity chart to 1.0 cm. Move directly across the chart until you cross the bold line. Drop straight down and the velocity will be a little under 100 cm/sec. Estimate it to be around 90 cm/sec.

10. **C** is **correct**. According to the chart, water with a velocity of 300 cm/sec can move the largest cobbles and of course everything smaller.

Answers Explained Lesson 5–6

1. **A** is incorrect because horizontal sorting comes from a gradual slowing of water velocity.

   B is **correct** because when the water suddenly stops, the sediments get dropped in size order with the largest and heaviest at the bottom.

   C is incorrect because the outside of meanders are erosional areas.

   D is incorrect because sub-soil is a layer of soil below the topsoil.

2. **A** is incorrect because the edges of a straight section of the stream have slow water because of friction between the water and the bank.

   B is incorrect because river deltas are where the stream slows down because it is entering a quiet body of water.

   C is **correct** because the fastest water gets thrown to the outside of the turns.

   D is incorrect because the deepest part of the stream is in contact with the stream floor and there is friction that slows the water down.

3. **A** is incorrect because the erosion on the outside of the turns and deposition on the inside cause the meanders to widen.

   B is **correct** because the outside of the meander will erode outward.

   C is incorrect because meandering rivers have more lateral erosion than down-cutting.

   D is incorrect because meandering rivers generally transport sand sediments and smaller.

4. **A** is **correct** because youthful streams have steep gradients.

   B is incorrect because the steep sides of the mountains will not allow a mature stream to begin to meander.

   C is incorrect because mountains have sides that are too steep for a slow meandering river.

   D is incorrect because old-aged streams do not happen in the mountains.

5. **A** is **correct** because the high energy of a youthful stream has a lot of down-cutting, which produces a v-shaped valley.

   B is incorrect because meanders are found in mature and old-aged streams.

   C is incorrect because u-shaped valleys are typical of glaciated areas.

   D is incorrect because mature streams have wide, eroded valleys.

6. **A** is incorrect because the source of the river is an erosional area.

   B is incorrect because, at the mouth of the river, the water velocity drops very low and deposition is dominant.

   C is incorrect because static equilibrium means that nothing moves. When erosion and deposition are balanced, some sediments come into the area and the same amount leave. This keeps the balance but things are moving.

   D is **correct** because a dynamic equilibrium describes a system in which two opposites work at the same but opposite rates.

7. **A** is **correct** because fast water gets thrown to the outside.

   B is incorrect because the slowest water is found at the inside of the curves.

   C is incorrect because the water is deepest near the outside curve of the stream.

   D is incorrect because the insides and outsides or curves will have the same clarity.

8. **A** is incorrect because vertical sorting is caused by a sudden stop in water velocity.

   B is **correct** because, as the water gradually slows, it will drop the largest particles still in transport.

   C is incorrect because meandering is a combination erosional and depositional feature.

   D is incorrect because cross-beds are formed from many events of gradual slowing.

9. **A** is incorrect because the round particles get deposited first.

   B is incorrect because round particles get deposited fastest.

   C is **correct** because flat materials flop and tumble around and get deposited slowly.

   D is incorrect because fast water will keep flat sediments suspended longer than quiet water.

Answers Explained Lesson 5–7

1. **D** is **correct** because mass wasting describes moving a massive amount of sediment that is pulled downhill by gravity.
2. **A** is **correct** because water is so common and falls from the sky. Water is also powerful.

   B is incorrect. Even though wind is just as common as water, wind does not have the power that water does.

   C is incorrect. Glaciers are extremely powerful, but they do not cover a large portion of the Earth.

   D is incorrect because only rocks that are situated on steep slopes will be affected by gravity working alone.
3. **A** is **correct**. Tiny pits rough up the surface of rocks just like a frosted window has been pitted.

   B is incorrect because layers will give a striped appearance to the rocks.

   C is incorrect. Crystals on a rock's surface will make it appear sparkly.

   D is incorrect because all rocks and sediments are made of minerals.
4. **A** is incorrect. A delta is deposit left at the mouth of a river.

   B is incorrect. A moraine is a deposit of unsorted sediment left by a glacier.

   C is incorrect. A meander is a bend in the stream.

   D is **correct**.
5. **A** is **correct**. It may also be a rockslide or a mudslide.

   B is incorrect because downhill creep is a slow process.

   C is incorrect because a meander is a bend in a stream.

   D is incorrect because dynamic equilibrium is a balance between two opposite processes.
6. **A** is incorrect because sharp and angular sediments come from a gravity or a glacial deposit.

   B is incorrect because small, frosted, and pitted surfaces come from being exposed to wind.

   C is **correct** because the rushing of the water tumbles the sediments into rounded smooth shapes.

   D is incorrect because water does not normally move very large sediments and will usually sort them.
7. **A** is **correct**. When you look at the cone of a delta it has the shape of a triangle just like the Greek letter delta.

   B is incorrect. Gamma is another Greek letter.

   C is incorrect. A meander is a curve somewhere in the path of the river.

D is incorrect because a cut bank is the outside bank of a curve where erosion takes place.

8. **C** is **correct**. Graded bedding is a stack of vertically sorted layers.

Answers Explained Chapter 5 Exam

Exercise A

1. **P** Scratching is an example of abrasion, a type of physical weathering.
2. **C** Dissolving is a chemical reaction.
3. **P** The pressure caused by the expansion of water is a physical force.
4. **C** Rusting comes from a chemical reaction with oxygen.
5. **P** The pressure caused by the growth of roots is a physical force.
6. **C** The discoloration of the Statue of Liberty is a chemical reaction.

Exercise B

7. **Sand** .07 cm is in between .006 cm and .2 cm.
8. **Silt** .005 cm is between .004cm and .006cm.
9. **Cobble** 20.5 cm is in between 6.4 cm and 25.6 cm.
10. **Pebble** 4.2 cm is in between .2 cm and 6.4 cm.
11. **Clay** .0002 cm is smaller than .0004 cm.
12. **Boulder** 39.5 cm is larger than 25.6 cm.

Exercise C

13. Pebbles of about .5 cm and everything smaller
14. Pebbles of about 2 cm and everything smaller
15. Cobbles of about 9 cm and everything smaller
16. Cobbles of about 20 cm and everything smaller
17. The smallest boulders and everything smaller

Multiple Choice

18. **A** is **correct**. Water expands and lowers its density when it freezes. This is why ice floats on top of liquid water.

    B is incorrect because water expands when it freezes.

    C is incorrect because water exerts tremendous pressure when it freezes and expands.

    D is incorrect because water turns solid when it freezes.

19. **A** is incorrect. Evaporated sea water gets into the air as pure water and all of the other substances in sea water are left behind in the ocean.

B is **correct** because carbon dioxide reacts with water to form carbonic acid (seltzer).

C is incorrect because the question asked for the natural source. Pollution is man-made (but it is a source of acid rain).

D is incorrect because when water gets into the air it is pure water. Any minerals that may have been dissolved in the water are left behind during evaporation.

20. **A** is incorrect. Unchewed candy has less surface area because your saliva can only work on the outside surface.

    B is incorrect. Chewing simply grinds the candy into smaller pieces.

    C is incorrect. Chewing is a physical action and does not change the candy's resistance to scratching. That would require a chemical change.

    D is **correct**. When you chew the candy it exposes the inside of the candy to your saliva.

21. **A** is incorrect. Physical weathering prefers cold or dry climates.

    B is **correct** because the warmth supplies energy and the moisture supplies a wet medium for chemical reactions to occur.

    C is incorrect because frost action needs moisture and a temperature that fluctuates above and below freezing.

    D is incorrect because wind abrasion does not work too well in moist climates. The water weighs down the sediments.

22. **A** is incorrect because the humus in topsoil makes it darker.

    B is incorrect. Topsoil and sub-soil have about the same size sediments. However, the topsoil is exposed to more intense weathering.

    C is **correct**. Topsoil and sub-soil are essentially the same except for the humus in the topsoil.

    D is incorrect because both top soil and sub-soil are made of mineral sediment.

23. **A** is incorrect. By definition, transported soil has been moved.

    B is **correct**. Transported soil is more common than residual soil because erosion is so common on the surface of the Earth.

    C is incorrect because both transported and residual soils have humus.

    D is incorrect because both types of soil can be found on river banks.

24. **A** is incorrect because the lower an object is, the lower its potential energy will be.

    B is incorrect because, as the sediments moves down a mountain, it will get weathered and worn into a smaller particle.

    C is **correct** because the lower an object is, the lower its potential energy will be.

D is incorrect. Recrystallization happens under the extreme heat and/or pressure of metamorphism.

25. **A** is incorrect because the stream did not get lower; it just tilted. A decrease in potential energy would require a lower elevation.

    B is incorrect because an increase in volume would require more water. Even though this stream will flow faster due to the steeper slope the amount of water in the river stays the same ("the faucet still has the same setting").

    C is incorrect because an increase in potential energy would require a higher elevation and the stream bed was tilted in place.

    D is **correct** because a steeper slope forces the water to slide downhill faster.

26. **A** is incorrect because oxidation refers to a chemical reaction between oxygen and another substance.

    B is incorrect because graded bedding is layers of vertical sorting. It is a depositional feature.

    C is incorrect because weathering is the breaking down of rocks. It does not involve transportation.

    D is **correct** because the transportation of rocks and sediments is erosion.

27. **A** is incorrect because minerals are the basic ingredients of all rocks.

    B is **correct**.

    C is incorrect because cross-beds are deposits.

    D is incorrect because deposits are made of many types of sediment.

Geologic History

Lesson 6–1: Relative Dating of Rocks

There are two ways to discuss the historical events of the Earth: in order of occurrence and by placing the event on a time line.

To date events by **relative time** is to place them in order of occurrence—which came first, and second, and so on. Relative time is a good way to discuss a sequence of events but gives no concept of how much time has passed. For example, the terms *World War I* and *World War II* indicate which came first, but from the sequence alone we have no information how much time passed between the two wars.

Absolute time assigns an exact date to an event such as "10,000 years ago" or "WWII started in 1942." Absolute time is more difficult to determine and often requires technology and lab work.

Relative time is often used to reconstruct the order of events that create a series of rock layers. In order to do so, a few rules and guidelines are used:

Uniformitarianism: This big word simply means that processes in the past happened the same way that they do today. In other words, they happened in a uniform manner. The concept of uniformitarianism is a simple and logical one—but it is an assumption. If we go to an active glacier and see how it works, what it does to rocks, and the types of deposits it leaves behind, and then when we see those same patterns in ancient rock, it is a safe assumption that a glacier was responsible for the formations. Uniformitarianism allows us to infer processes in the past without actually seeing them happen.

Uniformitarianism has a catch phrase: "the present is the key to the past." It means that the key to understanding how things happened in the past is to understand processes that we can observe today. For example, to understand how large creatures such as dinosaurs functioned, we study large creatures from today such as elephants, rhinos, whales, alligators, and so on.

Superposition: In a sequence of *undisturbed* sedimentary rocks, the youngest rocks will be at the top. When sedimentary rocks are made, they usually pile up one on top of the other similar to piling a stack of books one at a time. The first book on the pile, and therefore the oldest, will be on the bottom.

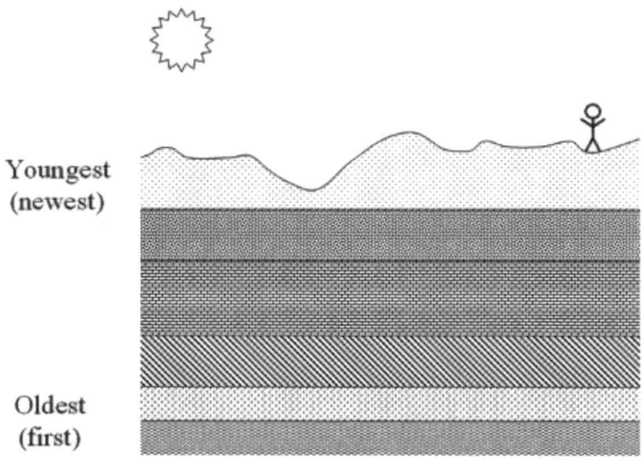

Youngest (newest)

Oldest (first)

Figure 6.1. The Law of Superposition states that the oldest layer is on the bottom.

There are two exceptions to the Law of Superposition. You can get older rocks on top of younger rocks if a section of rocks were overturned. This will happen if you have extensive folding of an area from mountain building. To see how this could happen, lay a piece of paper flat on a desk. Place one hand on each end of the paper and slowly push your hands together. The paper will start to fold. If one side gets pushed over (in this example, by one of your fingers), the paper will flop over to one side—placing the bottom side (the oldest) on the top side of the paper.

The other exception to superposition is an igneous intrusion. An igneous intrusion will squirt between preexisting layers. It will be younger than the two layers above and below it. The clue to look for here is contact metamorphism on both the top and bottom of the intrusion telling us that those layers were there first.

The youngest rocks will have the uppermost or "super" position, hence superposition.

Cross-Cutting Relationships: When any feature cuts across layers of rocks, the feature is younger than the rocks it cuts across. A common example of a cross-cutting relationship is an earthquake fault. In order for a fault to crack a rock or series of rock layers, the rock must exist first. Likewise, when magma melts its way towards the surface, it will intrude on the surrounding rocks. Therefore, an igneous intrusion is younger than the surrounding rocks that it cuts across.

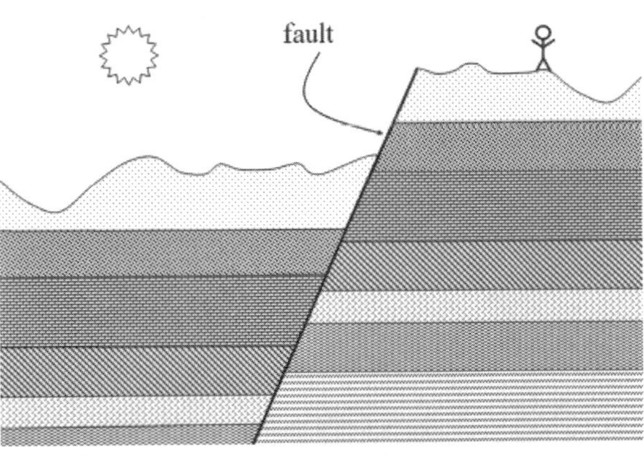

Figure 6.2. The Law of Cross-Cutting Relationships states that the layers were there first and then the fault came along after.

In the case of an igneous intrusion, contact metamorphism indicates that the surrounding rocks existed first and then got changed by coming into contact with extreme heat. If contact metamorphism is not present at the top of a lava flow, then it tell us that the layers above were deposited after the lava cooled.

Included fragments within a rock are older than the entire rock sample. A conglomerate will have three distinct formation events: the formation of the pebbles, the appearance of the mineral cement, and then the assembly of the pebbles to create the larger rock.

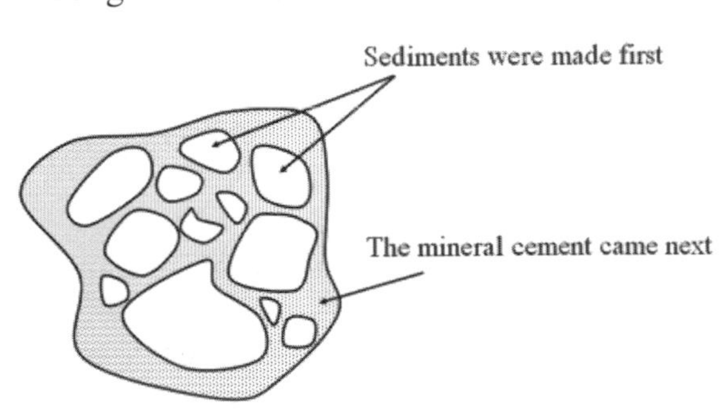

Figure 6.3. Included Fragments

One assumption that is usually safe to make is that sedimentary rocks are made in layers that are originally horizontal. This is called the **Principle of Original Horizontality**. There are exceptions to this generalization but they are specific circumstances such as the deposits in a river delta.

Folds or tilts in rocks are younger than the rocks themselves. This is an offshoot of the principle of original horizontality. Because most sedimentary rocks are made flat and level, if you see them in any other orientation, they were formed flat first and then changed. Typical changes will look like folded or tilted layers.

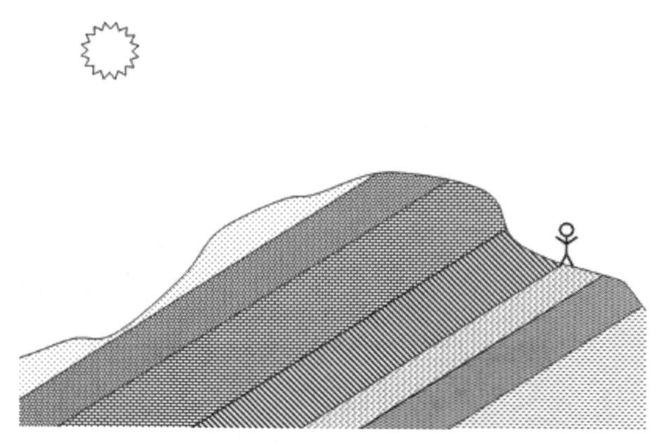

Figure 6.4. Tilts

There are other guidelines for figuring out the formation sequence of sedimentary layers:

Formation of sedimentary rocks usually happens under water. The vast majority of sedimentary rocks are deposited underwater. Unless there is other information available, it is a pretty safe assumption that sedimentary rocks indicate that the environment during formation was wet.

Weathering and erosion usually happen above water (on dry land). Weathering happens fastest wherever the rock material is exposed to the elements of nature.

An **unconformity** is a "buried erosional surface," which means that it was a surface that once got eroded and then buried by new layers. In order for an unconformity to form, three steps must occur: the area must be uplifted out of the water (where the previous layers were deposited), then eroded, then submerged again so that new layers can be placed on top. An unconformity represents a gap of history for which we have no information because we do not know how many layers were eroded.

Note: When you are doing the sequence puzzles at the end of this lesson, you need to identify the times when the environment was under water or exposed. The way that this happened are events called emergence and submergence. Emergence out of the water could be caused by either sea level dropping or the land being pushed upward. Submergence can be caused by sea level rising or the land sinking.

Correlation

Matching a layer in one location with a layer formed at the same time in another location is called **correlation**. When layers are correlated, we can match up the events and the fossils associated with the layers.

Correlation Techniques

"*Walking the outcrop*." An exposed layer of rock is called an **outcrop**. Walking the outcrop is following an outcrop and mapping all the structures that it touches.

Rocks can be correlated by matching similar rock characteristics, such as similar colors, compositions, a unique feature, or by the same series of rocks "sandwiching" it.

Using index fossils to find layers of the same age is one of the best ways of correlation. An **index fossil** is a fossil of an organism that existed in a wide area but for only a short time. It is important that all index fossils meet both of these requirements. This allows the geologist (or paleontologist) a good chance of finding the fossil and to be able to date the formation of the layer in which the fossil was found within a small range of dates. When an index fossil is found it is easy to narrow down the time range for the layer of rock in which it is found.

A **volcanic time marker** can be used to mark all the layers forming at the same time as a violent volcanic explosion. The eruption will throw dust into the atmosphere that will settle over a wide area in a short time. Due to the suddenness of deposition (less than weeks or months), volcanic time markers represent an instant in geological time. An asteroid impact can have the same effect such as the famous asteroid dust layer that marks the end of the appearance of dinosaur fossils 65 million years ago.

Lesson 6–1 Review

Exercise A

Put an R next to the statements that describe *relative time* and an A next to those that are *absolute time*.

1. _____ First came the bacteria, then the jellyfish, and then the fish.
2. _____ Sam is the third of five children.
3. _____ The first humans were here three million years ago.
4. _____ A volcanic eruption killed off the species.

5. _____ A volcanic eruption in 79AD destroyed Pompeii.

6. _____ Dinosaurs and humans never co existed. They were separated by a very long time.

Exercise B

7. Place the following events in their correct order, starting with the oldest event first. Note that not every event has a letter, and there are exactly enough spaces as are needed.

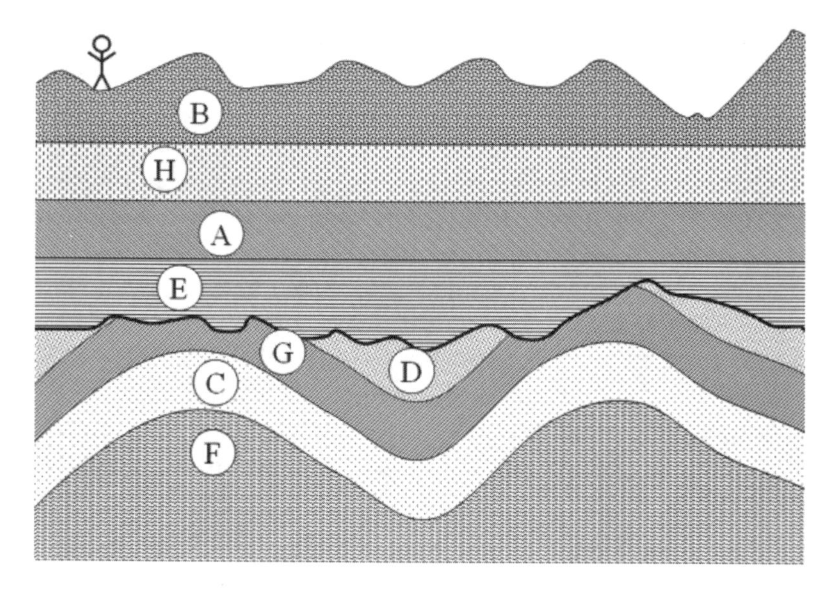

Figure 6.5

First 1) __________

2) __________

3) __________

4) __________

5) __________

6) __________

7) __________

8) __________

9) __________

10) __________

11) __________

12) __________

13) __________

Last 14) __________

Exercise C

8. Place the following events in their correct order, starting with the oldest event first. Note that not every event has a letter and there are exactly enough spaces as are needed.

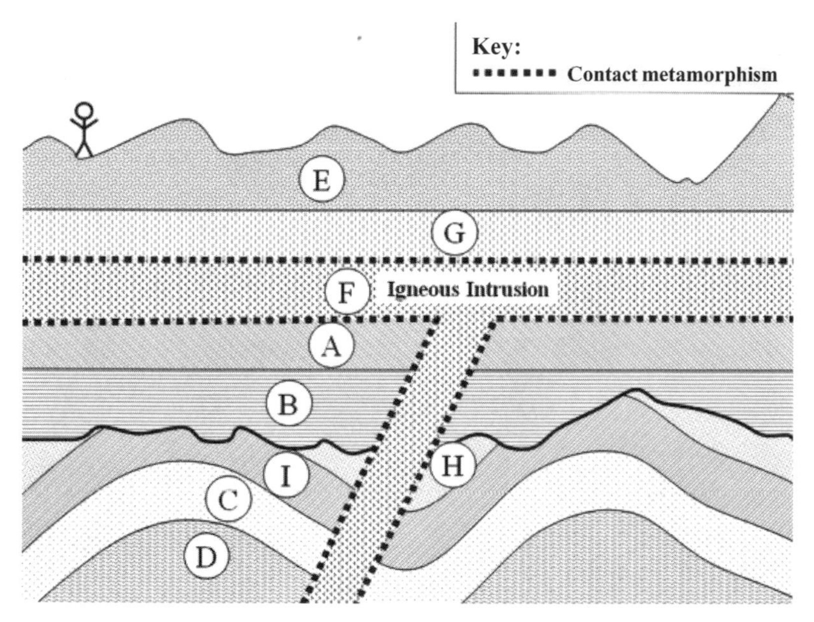

Figure 6.6

First 1) __________ 8) __________

2) __________ 9) __________

3) __________ 10) __________

4) __________ 11) __________

5) __________ 12) __________

6) __________ 13) __________

7) __________ 14) __________

Last 15) __________

Lesson 6–2: Absolute Dating of Rocks

Ways of Measuring Absolute Time

- **Tree rings**: Every year a tree grows a new ring. When the tree dies, the rings can be counted and events such as forest fires, rainfall, and disease can be dated. (Evidence of the same fire may be seen as a layer of ash in rocks to correlate a rock layer with the tree rings). Tree rings from fresh trees can be matched with ancient timbers in buildings. And the timbers in ancient buildings can be matched with the patterns in petrified (fossilized) wood. Careful matching of these different types of preserved wood can be used to date events up to thousands of years ago.

- **Varves**: Annual layers in sedimentary rock that were formed in a lake are varves. Some lakes have seasonal deposits which can be used as tree rings are to count back in time, for example, sand in the spring from high rain volume and silt deposited in the winter when the top is covered in ice and water flow stops.

Radioactive Dating

Here is some background information on radioactive elements.

An **atom** is the smallest piece of an element that still retains all of the element's properties. An atom is made of protons and neutrons in its center, or nucleus, and electrons "orbiting" around the outside of the nucleus. As far as radioactivity goes, the electrons are not part of this story.

The identity of an element comes from the number of protons within the nucleus. Hydrogen has one proton, carbon has six, and oxygen has eight. In order for a nucleus to remain stable, it also needs some neutrons in the nucleus. In addition, stable nuclei will usually have about the same numbers of protons and neutrons. If the numbers of protons and neutrons get too far out of balance the atom falls apart or "decays."

For example, carbon has six protons and six neutrons. However, there are other types of carbon atoms with more or less than six neutrons—which might upset the balance. All of these varieties of carbon, with differing numbers of neutrons, are called **isotopes**. The isotopes that have too much of an imbalance and decay are called **radioisotopes**.

When an atom undergoes radioactive decay, the nucleus will fly apart into pieces. Each piece takes some of the neutrons and protons. The number of protons attached to each piece will give the new identity to the decayed product.

The reason why radioactive decay works as a timer is that the decay happens at a steady rate and in a predictable manner. The amount of time that it takes for half of the atoms of a radioisotope to decay is called the **half-life**. For carbon-14, the half-life is about 5,700 years. However, after the second 5,700 years, there will be 25% of the parent material left (half of a half). If you were to track a radioisotope through its decaying life, it would have 50% after one half-life, 25% after two half-lives, 12.5% after three, 6.25% after four, and so on. Eventually, the decay would get down to the last atom. When that atom finally decays, all of the parent will be gone—you will *not* be left with a "half of an atom."

It is impossible to tell exactly which individual atom will decay but the decay rate holds true. Fifty percent will decay after one half-life. However, as the process continues and a small amount of the parent remains, the pattern will start to waver. To use an extreme example, when a half-life of carbon-14 starts with only three atoms, there cannot be exactly 50% remaining after 5,700 years because that would mean that there should be one and a half atoms remaining.

How does the decay rate of carbon-14 help tell how old an object is if you do not know how much carbon-14 the sample started with? When carbon-14 decays, it turns into nitrogen-14. Because the nitrogen-14 was formed from the decay of carbon-14, if you add all the C-14 and N-14 you will find out what the starting value of 100% was. For example, if a sample has 50 g of C-14 and 150 g of N-14,then the sample originally had 200 g of

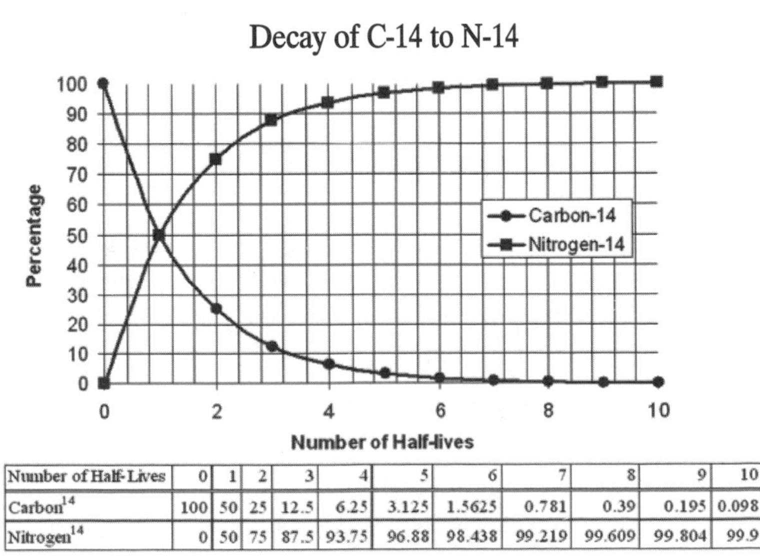

Number of Half-Lives	0	1	2	3	4	5	6	7	8	9	10
Carbon^{14}	100	50	25	12.5	6.25	3.125	1.5625	0.781	0.39	0.195	0.098
Nitrogen^{14}	0	50	75	87.5	93.75	96.88	98.438	99.219	99.609	99.804	99.9

Figure 6.7

carbon-14. A comparison of remaining 50 g of C-14 to the original total of 200 g will give a percentage of 25%, or two half-lives. Two half-lives for carbon-14 is 2 × 5,700 years or 15,400 years.

There are two limitations to using carbon-14 for dating materials: 1) It is only useful for organic materials, and 2) it is only effective for dating events that are relatively recent (in geological terms)—less than 50,000 years.

While an organism is alive it is constantly taking carbon into itself. Some of this carbon is carbon-14 and some is carbon-12. During its lifetime, the organism maintains an equilibrium between the isotopes in their bodies and their environment. The levels stay constant. When the organism dies, the levels begin to change and "the clock starts." As a result, the carbon-14 dating marks the moment of death of the organism. This is useful for dating an actual organism as well as artifacts such as cloth, wood, fur, and bone tools.

One drawback to using carbon-14 is that it can be twitchy. Ultimately, the amount of C-14 available in the atmosphere depends on the intensity of sunlight. The actual sun varies slightly in its output and therefore changes the amounts of C-14. The other radioactive isotopes discussed below have a much higher reliability and are often used to fine-tune the C-14 results.

Carbon-14 is not the only radioisotope that is useful for dating. There are dozens of useful radioisotopes of different elements. Each isotope has a different half-life, which makes it useful for dating the ages of different events. If an isotope with a short half-life is used to date a very old event, the parent material will decay too far to be useful. If the radioisotope decays with a very long half-life and the event is recent, then there will not be enough decay to measure and get a date. Carbon-14, with a half-life of 5,700 years, is useful for dating events that happened in the range of thousands of years ago. If the event is millions of years ago, Carbon-14 will have decayed completely and cannot be measured.

How to Calculate the Age of an Item Using Radioactive Isotopes

1. Add the amount of parent material with the daughter product. This will equal the total amount of material with which you started.

2. Find the percentage of parent material compared to the total material (parent/total × 100).

3. Find the number of half-lives by looking up the percentage on the carbon-14 decay chart (Figure 6.7).

Lesson 6–2 Review

1. If a sample contains 50 g of carbon-14 and 50 g of nitrogen-14, how many half-lives has it undergone?
   a) 1 b) 2 c) 3 d) 4
2. If a sample contains 25 g of carbon-14 and 75 g of nitrogen-14, how many half-lives has it undergone?
   a) 1 b) 2 c) 3 d) 4
3. If a sample contains 25 g of carbon-14 and 175 g of nitrogen-14, how many half-lives has it undergone?
   a) 1 b) 2 c) 3 d) 4
4. How old is a bone in which the carbon-14 in it has undergone 3 half-lives?
   a) 5,700 years b) 11,400 years c) 17,100 years d) 22,800 years
5. What percent of carbon-14 is left after 5 half-lives?
   a) 25% b) 12.5% c) 6.25% d) 3.125%
6. What happens to the amount of nitrogen-14 as the carbon-14 decays?
   a) The nitrogen-14 increases.
   b) The nitrogen-14 decreases.
   c) The nitrogen-14 remains the constant.
   d) The nitrogen-14 gets cut in half.
7. If a 20 g of carbon-14 has a half-life of 5,700 years, what would be the half-life of a 40 g sample?
   a) 2,859 years b) 5,700 years c) 11,400 years d) 20,000 years

Lesson 6–3: Ancient Life Preserved in the Rock

A **fossil** is any sort of evidence of past life on Earth. Technically, fossil means "dug up." Usually, it is the "hard parts" of an organism (teeth, bones, leaves, and twigs) that have been preserved. It is rare, but not unheard-of, to have fossils of worms and jellyfish or skin or internal organs. A fossil can also be evidence of life rather than the direct remains of it. A **trace fossil** could be the footprint of a walking creature, a worm burrow, or even droppings that have been preserved.

Evolution

Evolution, first proposed by Charles Darwin, is the changing pattern of life forms. The ultimate goal of nature and life is to survive long enough to reproduce. If an organism accomplishes this, then nothing else matters. Once the organism reproduces, its DNA has been passed on and it has fulfilled its obligation to nature.

The battle cry of evolution is "survival of the fittest." This phrase simply means that a species will survive as long as it is best suited for its environment. This is often called "natural selection," although the two terms do not mean exactly the same thing. If the organism is not fit to survive, it will have difficulty reproducing and nature will select it to go extinct.

A good illustration of evolution is the story of the giraffe. Millions of years ago, the giraffe's ancestors were the size and shape of a regular dog—just a four-legged animal with a tail, neck, and head. Imagine a herd of thousands of these pre-giraffes, all eating from the same bushes. Eventually, all of the leaves from head-level and beneath have been stripped and the herd starts to run short on food. Somewhere in the herd is an animal with a slightly longer neck—a genetic mutation that caused the deformity. Usually, a deformity will work against the animal's survival but in this case it helps: The animal can easily reach all of the food that is out of reach of the rest of the herd. The animal thrives, reproduces and passes its long-necked genes to its offspring. After doing this for many generations, all current giraffes have longer necks.

In retrospect, the giraffe's story can be confirmed by finding older and older fossils that retrace the natural history of the species. The gradual change in the species can be seen once the fossils are collected and laid side by side.

One of the unfortunate side effects of evolution is extinction. **Extinction** is the disappearance of a species forever. The most obvious cause of extinction is the lack of reproduction of a species—in other words, it dies out. The extinction could be caused by predators, disease, climate change, or a catastrophe. Extinction can also be the result of a "less negative" cause: The organism could have evolved. In this case, the original species would be gone, but the newest version would take its place. An example of this would be the extinction of *Homo erectus*. *Homo erectus* is now extinct because they have turned into us: *Homo sapiens*.

Abundance Charts

The following chart is an abundance chart. It shows a few species, when they existed, and how abundant each organism was. It reads as a sideways bar graph would, and the width of the line represents how many of the organism existed at that time. A wide line was a popular time. A narrow line was a time when there were very few of the organisms around. On this diagram, the species are drawn as simple to more complex shapes to represent evolutionary complexity that occurs as time passes.

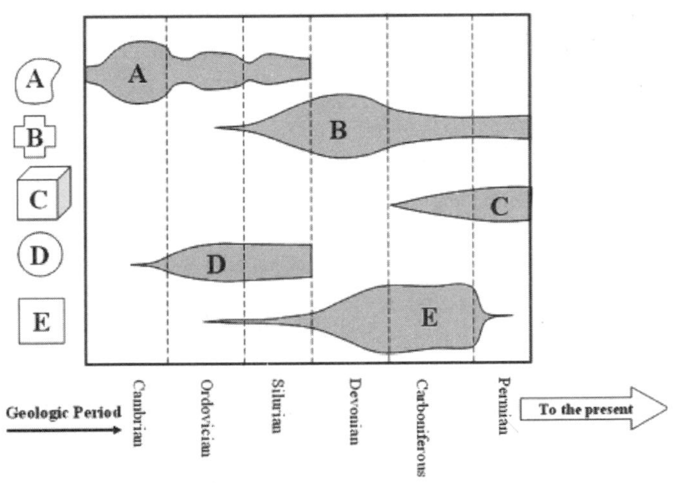

Figure 6.8. Abundance

Life on Earth

The Earth was formed about 4.6 billion years ago. Relatively soon after, about a billion years later, life began on the planet. This is rather quick considering the monumental step that it represents. From 3.5 billion years ago until about .5 billion years ago, the life on the planet was very simple. Most of life on the planet was made up of worms, slime, bacteria, and jellyfish. There were no hard-parts to preserve and fossils from before 500 million years ago are very rare. This is the time in the past called the Proterozoic Eon (*proto* = beginning, *zoic*, like "zoo," means life). The Proterozoic lasted for about 2 billion years with little diversity in life.

Then, about 544 million years ago (or for short, MYA), there was suddenly an explosion of life and diversification in the oceans. Fossils become abundant; there are shells and teeth to preserve. This explosion marks the beginning of the **Cambrian** period and is called the Cambrian Explosion. The Cambrian Explosion began the **Paleozioc**, or "old life," Era on Earth.

The oceans were teeming with all sorts of creatures that would look strange and alien to us today. Throughout the Paleozoic Era, life continued to flourish and diversify in the oceans and even began to colonize the barren land.

The Paleozoic Era ended with the arrival of the first dinosaurs. The age of the dinosaurs was the **Mesozoic** ("middle life") Era and it lasted from about 251 MYA until 65 MYA. The dinosaurs dominated the land. The first primitive mammals destined to evolve into humans hid under logs from larger predators. The mammals stayed in their submissive roles until the infamous end of the dinosaurs cleared out the top of the food chain—ending the Mesozoic Era.

The **Cenozoic** ("recent life") Era, sometimes called the age of mammals, began 65 MYA. During this time mammals diversified and took over most of the jobs that the dinosaurs did. Even as recent as a handful of millions or thousands of years ago, there were creatures, mammals, that looked strange and alien to us.

Mass Extinctions

A **mass extinction** is a period in the Earth's past where more species than normal went extinct. For reasons discussed earlier, organisms are constantly becoming extinct. But, there have been times in the past where a much greater than normal amount of species died off. Depending on where you wish to draw the line of "how bad it has before being to be called a mass extinction" there have been anywhere from five to 12 major disruptions of life on the planet. The most famous was the event that killed off the dinosaurs 65 MYA. But that wasn't the biggest. The extinction that ended the Paleozioc Era and made room for the dinosaurs 251 MYA wiped out more than 90% of all species on the planet! One out of 10 things survived!

The most recent trend in science is to blame most mass extinctions on the impacts of objects such as asteroids and comets. There has even been more and more evidence of this found using new technologies. When a large object hits the Earth, it throws up a tremendous amount of dust into the atmosphere as well as smoke from all the forest fires it ignites. The smoke and dust can be thick enough to block the Sun and cool the planet.

Asteroid and comet impacts are not the only culprit for mass extinctions. Some other theories that may have been responsible are:

- Disease.
- The arrival of a new predator.

- Changing climate—possibly from continents coming together or splitting apart.
- The arrival of humans.

Lesson 6–3 Review

Exercise A

Use Figure 6.8 on page 181 to answer questions 1–5.

1. During which geologic period did organism E go extinct?
   a) Silurian
   b) Devonian
   c) Carboniferous
   d) Permian
2. During which geologic period did organism B come into existence?
   a) Cambrian b) Ordovician c) Silurian d) Devonian
3. It appears that a sudden extinction event happened at the end of which period?
   a) Cambrian b) Ordovician c) Silurian d) Devonian
4. Which creatures coexisted with C?
   a) A and D b) A, B, and E c) B and E d) D and E
5. Place the creatures in their correct order of appearance.
   a) A D E B C b) A B D C E c) C B E D A d) A B C D E
6. Life first appeared on Earth during the
   a) Cenozoic Era.
   b) Mesozoic Era.
   c) Paleozoic Era.
   d) Proterozoic Eon.
7. The Cambrian Explosion was
   a) a major extinction event.
   b) the cause of Pangaea breaking up.
   c) a sudden diversification of life on Earth.
   d) a volcanic eruption that blocked out the Sun for years.
8. Fossils of jellyfish are rare because
   a) jellyfish do not have any hard parts.
   b) jellyfish are rare.
   c) they are buried too deep.
   d) they are only found in remote places on the Earth.

Chapter Exam

Exercise A

1. Place the following events in their correct order, starting with the oldest event first. Note that not every event has a letter and all spaces are needed.

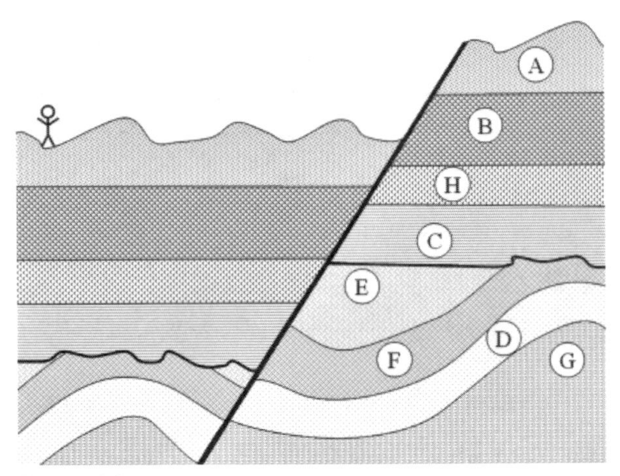

Figure 6.9

First 1) ________

2) ________

3) ________

4) ________

5) ________

6) ________

7) ________

8) ________

9) ________

10) ________

11) ________

12) ________

13) ________

14) ________

Last 15) ________

Exercise B

Use the following figures to answer questions 2 and 3.

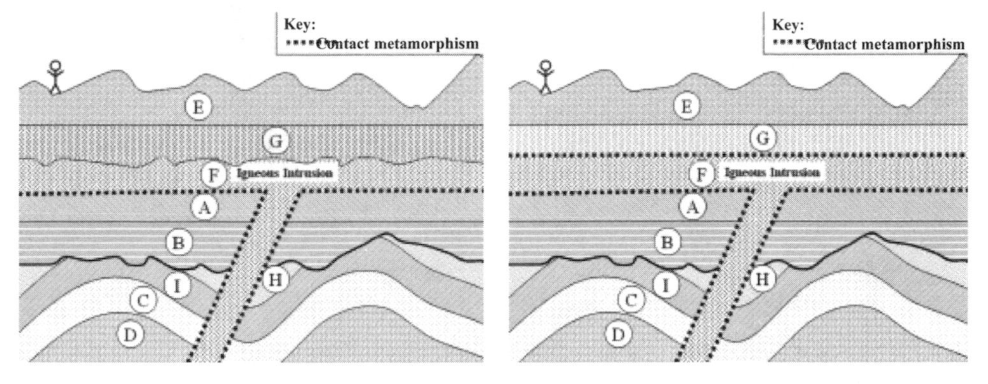

Figure 6.10

Figure 6.6

2. There is one difference between the two drawings. Identify the difference.
3. Why is the difference important?
4. Place the events for the sequence shown in figure 6.10 in their correct order, starting with the oldest event first. Note: the sequence is not exactly the same as Exercise C of Lesson 6–1.

First	1) ________		10)	________
	2) ________		11)	________
	3) ________		12)	________
	4) ________		13)	________
	5) ________		14)	________
	6) ________		15)	________
	7) ________		16)	________
	8) ________		17)	________
	9) ________	Last	18)	________

5. Which of the following will make a good index fossil to people of the future?
   a) Mountain lion, because it existed for 10 million years and was sparsely populated in the western United States.
   b) Horseshoe crab, because it lived throughout the oceans and has been unchanged for hundreds of millions of years.
   c) Coelacanth, because it is a lobe-finned fish so rare that it was thought to be extinct until a recent discovery. It has existed since before the dinosaurs.
   d) Humans, because they colonized the entire planet and existed for three million years.
6. An exception to the Law of Superposition could be
   a) folded layers.
   b) sandstone on top of limestone.
   c) a lava flow on the surface.
   d) metamorphic rock on the bottom.

Use the following figure for question 7, which shows a close-up view of a clastic sedimentary rock.

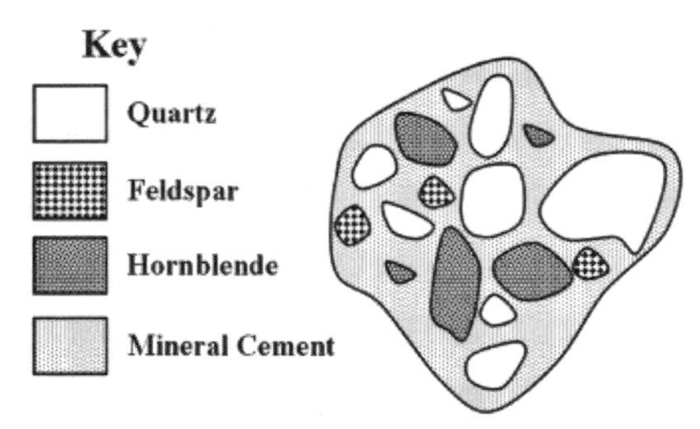

Figure 6.11

7. In Figure 6.11, which was the most recent event?
   a) Formation of the quartz
   b) Weathering of the hornblende to its current shape
   c) Accumulation of the mineral cement
   d) The crystallization of hornblende inside a volcano

8. After one half-life, 50% of a radioactive element will remain. How much will be left after two half-lives?

   a) 50% b) 25% c) 4% d) 75%

9. Carbon-14 is

   a) the only radioactive element that is useful for radioactive dating.
   b) used to date really old rocks.
   c) useful because it never decays totally.
   d) used to date recent (less than 50,000 years) organic remains.

10. Daughter products

    a) are more radioactive than parent material.
    b) increase over time.
    c) are only found in metamorphic rocks.
    d) are always heavier than the parent material.

11. If you add the amount of daughter product and the parent material, you should always get

    a) 100%.
    b) 50%.
    c) 0%.
    d) different percentages.

12. The slowly changing pattern of life is called

    a) extinction.
    b) uniformitarianism.
    c) evolution.
    d) dynamic equilibrium.

13. Dinosaurs lived during the

    a) Proterozoic Era.
    b) Paleozoic Era.
    c) Mesozoic Era.
    d) Cenozoic Era.

Answer Key

Answers Explained Lesson 6–1

Exercise A

1. **R** You know the order, but not how much time has passed.
2. **R** You know some of the order, but not how much time is in between each child. They could be quintuplets.
3. **A** You can find three million years ago on a time line.

4. **R** You know the species was around first and then the eruption right at the end, but you don't know when it happened.
5. **A** You can place 79AD on a time line.
6. **R** You know the dinosaurs came first and are now gone, but this statement has no information about how much time has passed. You cannot place it on a time line.

Exercise B

7.
1) **Layer F is deposited** (flat and horizontal).
2) **Layer C is deposited.**
3) **Layer G is deposited.**
4) **Layer D is deposited.**
5) **Folding** (all of the layers together).
6) **Uplift** (all of the layers together).

Steps **5 and 6** can be switched or can happen at the same time.

7) **Erosion.**
8) **Submergence.**
9) **Layer E is deposited.**
10) **Layer A is deposited.**
11) **Layer H is deposited.**
12) **Layer B is deposited.**
13) **Uplift.**
14) **Erosion.**

Exercise C

8.
1) **Layer D is deposited.**
2) **Layer C is deposited.**
3) **Layer I is deposited.**
4) **Layer H is deposited.**
5) **Folding** (all of the layers together).
6) **Uplift** (all of the layers together).

Steps **5 and 6** can be switched or can happen at the same time.

7) **Erosion.**
8) **Submergence.**
9) **Layer B is deposited.**
10) **Layer A is deposited.**
11) **Layer G is deposited.**

Anytime after this point, the intrusion of **F** can happen.

12) **Intrusion of F and contact metamorphism.**

The **contact metamorphism** happens at exactly the same time as **F** and is part of the same step.

13) **Layer E is deposited.**

14) **Uplift.**

15) **Erosion.**

Answers Explained Lesson 6–2

1. **A** is **correct**. When there is the same amount of carbon-14 as nitrogen-14, it means that 50% of the carbon-14 has turned into nitrogen-14. It has been "cut in half" once: one half-life.
2. **B** is **correct**. Start with 100% carbon-14 and count backwards by half counting on your fingers each time you take away half. 100% = 0; 50% = 1; 25% = 2 half-lives.
3. **C** is **correct**. This problem is a little trickier because you don't start with 100 grams = 100%. One hundred percent in this problem is 200 g (add 25 g + 175 g). If you start with 200 g as your original parent material, 25 g is 12.5%. Now count backwards:
   100% = 0; 50% = 1; 25% = 2; 12.5% = 3 half-lives.
4. **A** is incorrect. 5,700 years is one half-life.

   B is incorrect. 11,400 years is two half-lives.

   C is **correct**. 3 × 5,700 years is 17,100 years.

   D is incorrect. 22,800 years is four half-lives.
5. Start with 100% carbon-14 and count backwards by half-counting on your fingers each time you take away half.
   100% = 0; 50% = 1; 25% = 2 half-lives.

   A is incorrect because 25% is two half-lives.

   B is incorrect because 12.5% is three half-lives.

   C is incorrect because 6.25% is four half-lives.

   D is **correct** because 3.125% is five half-lives.
6. **A** is **correct** because the nitrogen-14 is made as the carbon-14 decays.
7. **B** is **correct** because the half-life of carbon-14 always remains constant.

   A, C, and **D** are incorrect because the half-life of carbon-14 remains constant no matter how much of a sample you have.

Answers Explained Lesson 6–3

1. **D** is **correct** because the last appearance of organism E was during the Permian.

2. **B is correct** because the first appearance of organism B was during the Ordovician.
3. **C is correct** because two different species suddenly died off at the end of the Silurian.
4. **A** is incorrect because A and D disappeared long before C appeared.

   B is incorrect because A disappeared long before C appeared.

   C is **correct** because B and E overlap with C.

   D is incorrect because D disappeared long before C appeared.
5. **A is correct** because A arrived first, then D, E, B, and finally C. If you had difficulty with B and E, look at the pictures of the organisms on the left of the diagram. E is simpler (less corners) than B, which suggests that E came first.
6. **D is correct**. Proterozoic means "beginning life"
7. **C is correct**. At the beginning of the Cambrian, life suddenly "exploded" throughout the oceans of the planet.
8. **A is correct**. The best preserved parts of an organism are the hard parts such as bones and teeth.

Chapter 6 Exam Answers Explained

Exercise A

1.
   1) **Layer G is deposited.**
   2) **Layer D is deposited.**
   3) **Layer F is deposited.**
   4) **Layer E is deposited.**
   5) **Folding** (all of the layers together).
   6) **Uplift** (all of the layers together) .

   Steps 5 and 6 can be switched or can happen at the same time.

   7) **Erosion.**
   8) **Submergence.**
   9) **Layer C is deposited.**
   10) **Layer H is deposited.**
   11) **Layer B is deposited.**
   12) **Layer A is deposited** .
   13) **Uplift.**
   14) **Erosion.**
   15) **Faulting** (of all the layers).

Exercise B

2. The difference between the two drawings is that this drawing (Figure. 6.10) is missing contact metamorphism on the top of Layer F.

3. The missing contact metamorphism tell us that the above layers were deposited after the lava cooled. In the previous version, Layer F was inserted between two layers. In this example, the above layers are younger than Layer F.

4. First
   1) **Layer D is deposited.**
   2) **Layer C is deposited.**
   3) **Layer I is deposited.**
   4) **Layer H is deposited.**
   5) **Folding** (all of the layers together).
   6) **Uplift** (all of the layers together).

   Steps 5 and 6 can be switched or can happen at the same time.

   7) **Erosion.**
   8) **Submergence.**
   9) **Layer B is deposited.**
   10) **Layer A is deposited.**
   11) **Layer F is deposited.**
   12) **Uplift** (all of the layers together).
   13) **Erosion.**
   14) **Submergence.**
   15) **Layer G is deposited** .
   16) **Layer E is deposited** .
   17) **Uplift.**
   18) **Erosion.**

   Note: The lack of contact metamorphism on the top of Layer F tells us that it was either a surface lava flow or it intruded between two layers and all of the overlying layers were eroded before Layer G was deposited.

5. **A** is incorrect because sparsely populated mountain lions will make the future fossil very rare. Index fossils need to be fairly abundant.

   B is incorrect because if a horseshoe crab fossil is found, it will be very difficult to tell it apart from one that is hundreds of millions of years old.

   C is incorrect because rare creatures do not leave enough fossils behind to be useful for correlation (matching rocks in distant areas).

D is **correct** because humans are abundant and widespread—an ideal index fossil. (We also bury our dead, which will improve the chances of fossilization.)

6. **A** is **correct** because layers that are folded could get overturned, which will place the older rocks on top of the younger.

   B is incorrect because sandstone and limestone can be formed at anytime. Neither one needs to be first.

   C is incorrect because a lava flow on the surface still has the youngest rock (the solidified lava) on top of the older rock.

   D is incorrect because metamorphic rocks can be old or recent. The type of rock does not help us with relative age.

7. **A** is incorrect because the quartz was originally formed before it became part of the rock shown.

   B is incorrect because the hornblende was weathered before it was assembled into this rock.

   C is **correct** because the mineral cement glued the preexisting sediments together to make this rock.

   D is incorrect because the hornblende was formed before it was assembled into this rock.

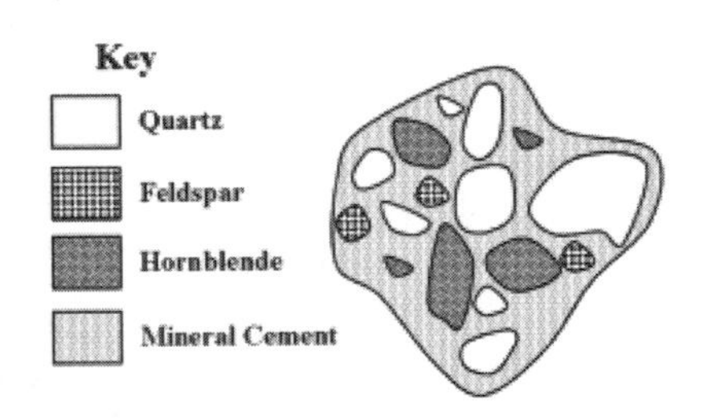

Figure 6.11

8. **A** is incorrect because each half-life removes half of what was present at the beginning of that half-life.

   B is **correct** because half of 50% is 25%.

   C is incorrect.

   D is incorrect.

9. **A** is incorrect because there are other elements that are useful for radioactive dating.

   B is incorrect because carbon-14 is only used to date materials that are less than 50,000 years old.

   C is incorrect because it eventually will decay completely. After about 50,000 years it has decayed too much to be useful.

   D is **correct** because carbon-14 is only useful for organic remains and is no longer useful after 50,000 years.

10. **A** is incorrect because daughter products can be either stable or radioactive.

B is **correct** because the daughter product forms as the parent decays. As more parent decays, more daughter forms.

C is incorrect because daughter products can be found in any type of rock.

D is incorrect because daughter products are often lighter than the parent material because some material has flown away.

11. **A** is **correct** because the daughter forms from the parent. The parent starts at 100% and the amount of the parent that decays turns into the daughter product—keeping the total at 100%.

12. **A** is incorrect because extinction is the end of a pattern of life.

    B is incorrect because uniformitarianism is a description of past geological processes.

    C is **correct**.

    D is incorrect because dynamic equilibrium is a balance between two opposite processes.

13. **A** is incorrect because the Proterozoic Era was the time of the first life on Earth—very simple organisms.

    B is incorrect because the Paleozoic Era was the time of the first abundant life—mostly sea life with shells.

    C is **correct** because the dinosaurs of the Mesozoic Era are the "middle" life between the primitives and the more recent creatures.

    D is incorrect because the Cenozoic Era is the recent life that took over after the dinosaurs went extinct.

Meteorology and Energy in the Atmosphere

Lesson 7–1: Structure of the Atmosphere

Meteorology is the study of the atmosphere—including the weather. **Weather** is the condition of the atmosphere. It can change from time to time and place to place. **Climate** is the type of weather an area has over a long period of time.

Structure of the Atmosphere

The majority of the atmosphere is near the Earth's surface because it is held by gravity. The atmosphere gradually fades away at the upper levels.

The atmosphere is divided into layers as see in Figure 7.1 based on temperature patterns. To observe these patterns, all one has to do is to travel straight up and graph the temperature. The pattern as shown in Figure 7.2 will emerge. The graph shows that it gets colder as you go up through the troposphere. You will know that you've reached the tropopause because the temperature will stop ("pause") getting colder. Once you get above the tropopause and into the stratosphere, the temperature will start to get

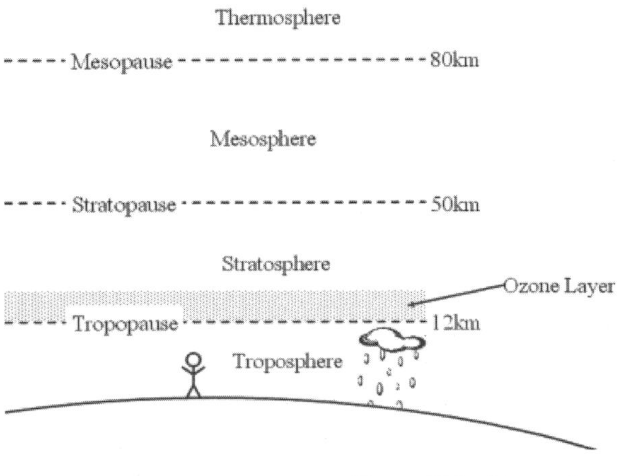

Figure 7.1. Atmosphere

hotter. This is caused by sunlight reacting with ozone in the stratosphere. It will continue to get hotter as you get higher until you reach the stratopause.

Troposphere

The bottommost layer of the atmosphere, and the one in which we live, is the **troposphere**. This is the layer that contains all of the weather that affects the surface or the Earth. This layer is identified by its temperature pattern of colder temperatures with increased height, and it extends up to about 18 km. This explains why the tops of mountains are cold and can have snow caps at higher elevations.

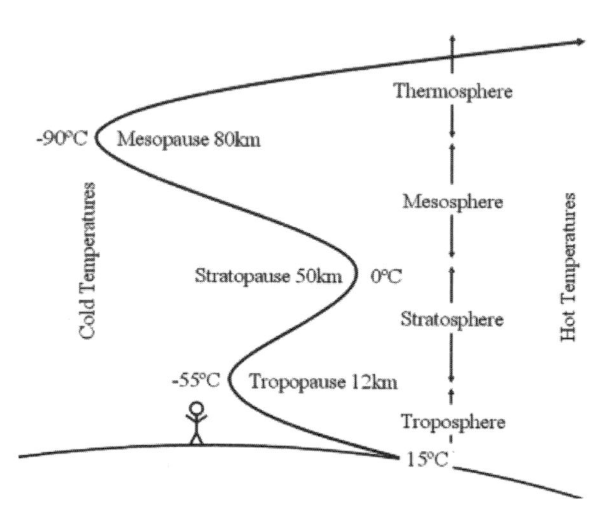

Figure 7.2. Temperature patterns

One reason why the troposphere has all of the weather is that it is the only layer with an appreciable amount of water vapor. When water evaporates from the surface, it has a hard time getting any higher than the top of the troposphere.

Stratosphere

The stratosphere is probably the most famous atmospheric layer because of two major events that occur here. Passenger jets have their cruising altitude within the stratosphere, and it is also the home of the ozone layer.

The **ozone layer** is a small section of the stratosphere with elevated levels of ozone (O_3). Ozone's main function in this layer is to catch the harmful ultraviolet rays of sunlight. When the reaction between sunlight and ozone occurs, the O_3 is changed into O_2. Ozone is naturally produced in the path of lightning strikes and has a "tangy" smell to it. You can smell the freshly produced ozone after taking off a sweater that has a lot of static in it. Because ozone is highly reactive, it is harmful to the lining of our lungs when breathed in. Occasionally, usually during the summer, large cities will have "ozone alerts." Pollution, coupled with strong summer sunlight, will produce ground level ozone, which is harmful to us and does not help thicken the ozone layer high above.

There are other layers of the atmosphere (the mesosphere, thermosphere, and the highest layer, the **exosphere**) but they have little impact on our daily weather.

Composition of the Atmosphere

The atmosphere is a mixture of gases, but it is mostly made of nitrogen (78%) and oxygen (21%). All the other gasses in the air are in very small, trace amounts. Despite their small concentrations, some gases have a huge effect. Some notable gasses are the greenhouse gases: carbon dioxide (CO_2) and methane; ozone (O_3), and water vapor (H_2O).

We, as oxygen-breathing animals, get the sense that the air is mostly oxygen, but oxygen is only a small portion of the air we breathe. The level of oxygen in the atmosphere is currently at 21%. This value has adjusted throughout history through a balance between the amounts of plant and animal life on the planet.

Lesson 7–1 Review

1. Weather occurs
   a) only in the troposphere.
   b) in the troposphere and stratosphere.
   c) in the stratosphere.
   d) above the stratosphere.

2. The condition of the atmosphere at a given location for a certain short period of time is called
   a) the watch.
   b) weather.
   c) the relative humidity.
   d) a contingency table.

3. The atmosphere is made mostly of the gas
   a) oxygen.
   b) carbon dioxide.
   c) nitrogen.
   d) methane.

4. The greenhouse effect is mainly caused by elevated levels of
   a) ozone.
   b) oxygen.
   c) carbon dioxide.
   d) chlorophyll.

Lesson 7–2: Atmospheric Temperature

Heat is the amount of energy an atom has due to its motion. Atoms with a lot of heat will move quickly or vibrate rapidly in place. Colder atoms will not have as much motion.

One of nature's goals is to "even things out." Nature hates imbalances and therefore tries to keep everything at the same heat level. As a result, if there is a difference in heat levels from one place to another, the heat will always migrate from the higher "source" to the lower "sink."

The heat content of a material is measured in calories. One **calorie** is the amount of heat needed to raise the temperature of 1 gram of water by 1°C. The "calories" that are used for dieting are called "kilocalories" and represent the amount of heat energy needed to raise 1,000 g of water 1°C.

Specific Heat

Not all substances react the same way when heat is put into them. Some materials are easy to heat up. In other words, these materials change their temperature greatly with very few calories added. On the other hand, there are some materials that will barely change their temperature even after a large amount of heat is added. **Specific heat** is a property of a substance that tells how much hotter it will get after adding heat.

Once again, water is used as the basis for the specific heat scale. Water has a specific heat of 1, which means that 1 gram of water will change its temperature by 1°C after 1 calorie of heat is added. As far as specific heats are concerned, this is a very high number and means that it is very difficult to change water's temperature. The difficulty with which water, or any material, heats up also applies when the material cools off. If it is difficult to heat up, it will also be difficult to cool down. This has huge implications for coastal versus inland climates.

Water is the Strangest Thing on Earth!

Water is the most common substance on the planet so we are very familiar with it. Despite this, water is the exception to almost every rule out there.

- It expands when freezing.
- Its solid is less dense than its liquid phase, and it floats.
- It is the universal solvent.
- It has a very high specific heat. It is very difficult to heat up and cool down.

A simple demonstration that shows the effect of different specific heats would be to heat equal amounts of different materials in boiling water and then placing them on slabs of wax to see which melts deeper into the wax. The deeper the melting, the more difficult the material is to cool off and therefore the higher the specific heat.

Heat and Phase Changes

Normally, when heat is added to a material, the material would get hotter. However, this does not happen when the material goes through a phase change. During a phase change, the added heat is used to break the bonds that hold the molecules together in a solid or a liquid. In the case of water, the phase change from solid to liquid (which is usually called melting) requires 80 calories of heat for every gram. While the 80 calories of heat are being added during the phase change, the temperature of the water (ice) will not go up until all of the ice is melted. Likewise, when water is boiling, the temperature will not go up until all of the water boils completely into a gas. To boil into gas, liquid water requires 540 calories of heat for every gram.

The really strange thing here is that during a phase change, the temperature will not change—no matter how hot the flame is under the water being heated.

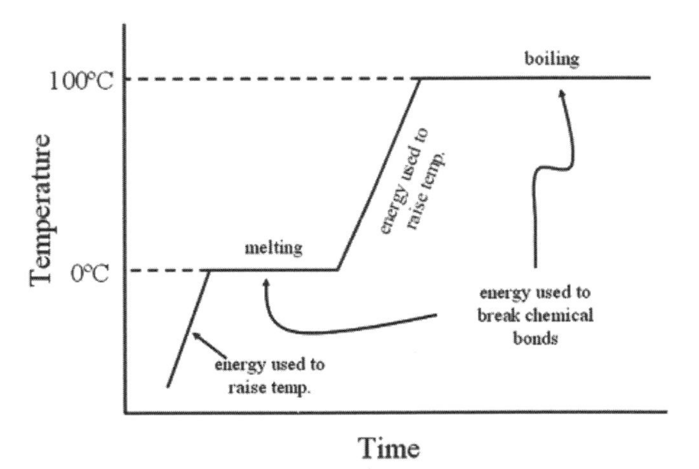

Figure 7.3. This graph shows the temperature of a sample of water that is being heated over a constant source of heat. Even though heat is constantly being pumped into the water, the temperature does not always rise, because during a phase change heat is used only to break the chemical bonds.

Heat Transfers

Radiation

Radiation is the transfer of heat in the form of light. Radiation is the only heat transfer method that can travel through the emptiness of space. The vast majority of the energy that drives our weather comes from the sun by way of radiation. (Any heat escaping from inside Earth is negligible to the planet's weather systems.)

There are two kinds of radiation: electromagnetic and ionizing.

Electromagnetic radiation (EM radiation) is a fancy way of describing the many forms of light such as yellow, red, radio, ultraviolet, and so on. **Ionizing radiation** comes from the dangerous particles that fly out of radioactive materials.

When the sun's light reaches Earth, it first penetrates the atmosphere. During its trip through the air, some of the light may get absorbed by any molecules or particles in the air. However, most of the energy reaches the surface. When the light strikes the surface, most of the energy gets absorbed and heats the surface.

The amount of energy that gets absorbed by the surface depends on the *color* and the *type* of surface. A lightly colored surface will reflect much of the light upwards into the sky and ultimately back into space. A good example of this would be the polar ice caps and the white tops of clouds. Dark materials such as dark rock or forests will absorb more energy.

The type of surface will affect how much light gets reflected. A smooth, shiny surface will act as a mirror and reflect some of the light, regardless of how light or dark it is. A rough, uneven surface will trap more of the light. **Albedo** is a term that describes a surface's reflectivity.

When a material absorbs heat from radiation, it must eventually release the heat. Otherwise, the material will continue to heat up forever. When a surface emits heat after absorption, the heat will escape as radiated infrared light. Any material that is a good absorber of heat must also be a good radiator. Likewise, "bad" absorbers are bad radiators.

Most of the energy emitted by the Sun is radiated as visible, shortwave light. This light gets absorbed by surfaces and turned into heat. At night, when the objects begin to cool down, the heat is reradiated as infrared, long wave light.

The Greenhouse Effect

The natural state of the Earth is to reradiate just as much heat as enters the Earth's system. This **dynamic equilibrium** between heating up in the warmth of the day and cooling down at night maintains the Earth's average temperature at a constant level. If that were not the case, a small imbalance, no matter how small, would add up through the millions of years the Earth has existed and cause the temperatures to plummet or skyrocket.

Considering Earth's distance from the sun, the reflectivity of the oceans, and the ice caps, our planet should be many degrees colder. However, gases in the atmosphere trap some of the infrared light that is reradiated back towards space. The result is an elevated temperature at which the balance between heating and cooling takes place—in other words: the world is warmer than it should be.

The gasses responsible act as the clear glass in a greenhouse that allows the sun's light to penetrate, yet stops the heat from escaping back out. For this reason, the insulating property of the atmosphere is referred to as **the greenhouse effect**. The main gases responsible for Earth's greenhouse effect (greenhouse gases) are carbon dioxide (CO_2), methane, and water vapor.

The major sources of CO_2 are organisms exhaling, forest fires, and burning of fossil fuels. In addition, the clear-cutting of forest land prevents trees from removing CO_2 from the atmosphere. Methane is produced from rotting vegetation and mainly comes from swamps and farting cows. Water vapor comes from evaporation of surface water, as the world gets hotter, more water will evaporate into the air.

Human activity since the Industrial Revolution has caused the levels of greenhouse gasses to increase over the last 150 years. As a result, the atmosphere is trapping more heat than the natural amount. The world is slowly getting warmer.

Conduction

Conduction is the transference of heat from one object to another through direct contact. A common example of conduction is burning your hand by touching a hot pan. Atoms with a lot of heat will move and vibrate faster. When they collide with other atoms, some of the kinetic energy of motion will transfer into the other atoms—making them move faster and feel hotter. The two main ways that conduction works in Earth Science is between the contact of hot magma and the surrounding rocks and also between the hot, sunlit ground and the air above it.

Convection

Convection can be described using only four words:

hot rises **cold sinks**

Technically speaking, **convection** is the circulation of heat caused by hotter air (or other substance) rising and colder air sinking due to density differences. Most materials will expand when heated. This will reduce the density, making the material lighter and it will want to float. Conversely, cooling a material will cause its atoms to get closer, making the material denser and it will want to sink.

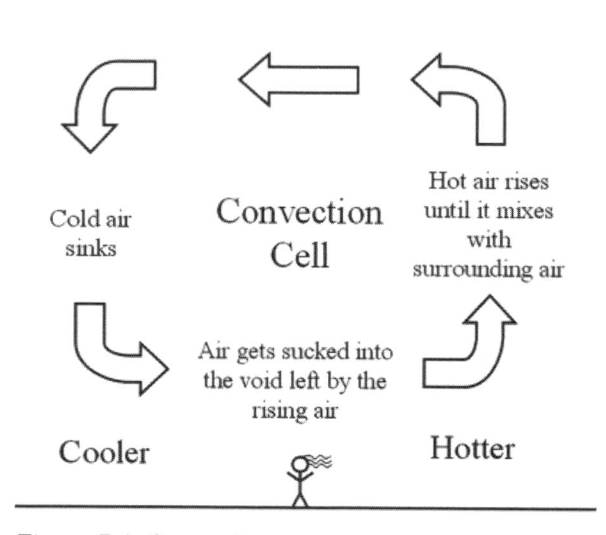

Figure 7.4. Convection

The pattern of hotter material rising and cooler material sinking is a **convection current**, and the looping cycle is a **convection cell**. In actuality, the hot material does not need to be "hot." It simply needs to be hotter than its surroundings. The same goes for the colder half of the cycle. For example, convection currents drive the inner workings of the Earth. The cooler, sinking currents within the Earth are by no stretch of the human imagination cold. The cooler currents are thousands of degrees, but they are still cooler than the surrounding hot, rising sections!

If you wanted to point your finger at the one thing that caused the most activity on the planet, it would probably be convection. Convection drives the currents within the mantle and lower crust, which causes the continents to drift, earthquakes, and the collisions that form mountains. Convection churns the atmosphere creating all of the weather. Convection circulates the ocean waters around the planet.

The three methods of heat transfer work as a team to distribute energy around the world. Radiation takes heat from its source, the sun, and brings it to Earth. The ground, heated by solar radiation, gives its heat to the air that it touches through conduction. Finally, the hot air rises and begins the cycle of convection. Ultimately, convection takes heat from the equatorial regions where radiation is most intense and brings it towards the poles. Nature does its best to balance everything in the middle. The goal of weather is to equalize the temperature of the planet.

If Hot Rises and Cold Sinks, Then Why Does Ice Float?

For most conditions, water will obey the general rule that hotter material will get less dense and float upwards while the colder stuff sinks. However, water does a strange and unique thing while it freezes: It expands. It is just about the only material on the planet that does so. As water starts freezing, the molecules of the liquid start to rearrange into the pattern of ice. The molecular pattern of ice takes a little more room than the liquid pattern. The arrangement of liquid water molecules is like clasping your hands by slipping your fingers between each other. When the water solidifies, it would be like spreading your hands apart so that each fingertip touches its mate on the other hand.

This rearrangement begins around 4°C and the expansion slowly continues until the water is completely solid. As a result, the densest water has a temperature of 4°C. Water that is 3°C, 2°C, and 1°C has already started going through the expansion.

Just about all of the liquids we are familiar with are water-based: soda, milk, even beer and wine. As a result, the expansion of water is commonplace but does not happen with any other material that goes through the liquid/solid phase change.

This unique property has had a critical impact on life here on Earth. When a lake freezes, the less-dense ice forms on the top, leaving liquid water below it. The remaining liquid is insulated by the cap of ice and doesn't always freeze. If water turned solid like any other substance, the ice would form on the bottom and the lake would quickly turn completely solid. This would kill most or all life in the lake.

Temperature

All matter has some motion such as vibration, spinning, or moving from place to place. Kinetic energy is the energy of motion. The more motion an atom has, the more kinetic energy it will have. When an atom is moving fast, it will have a lot of kinetic energy. If it bounces against your skin, it will feel hot. If it is moving slowly when it hits you, it will feel cool. In any material, the atoms will not have exactly the same amount of motion. When the temperature of a material is measured, the thermometer is measuring the average kinetic energy of the atoms touching the thermometer.

To help visualize, imagine two bowls of water, one hot and one cold. When you put your hand in the cold water, the atoms are moving slowly and do not bounce as hard against your hand—it feels cold. Place your hand in the hot water and the rapidly moving atoms hit against your skin hard and often—transferring more energy into your hand, and you feel the water as hot. If the two waters are mixed, the "hot" atoms are still moving fast and the "cold" atoms are still moving slow. However, when you place your hand in the water with the fast and slow atoms bouncing into you, it will feel warm. Your skin instantly registers the hard hits, the slow hits, and how often it is hit. The result: Your skin takes the average of all the energy being transferred and you feel warm.

There are three scales to measure temperature: Fahrenheit, Celsius (also known as Centigrade), and Kelvin. They all measure the average kinetic energy of a material, but they use different numbers to represent the temperature.

Fahrenheit

The Fahrenheit scale, invented by G. D. Fahrenheit, was based on two temperatures: a mixture of ice and salt that was the coldest temperature he could create in the lab and human body heat. He assigned zero as the coldest (0°F) and 96°F as human body heat. The Fahrenheit scale is used in the United States (including American meteorology).

In the Fahrenheit scale:

- Water freezes at 32°F.
- Water boils at 212°F.

Celsius

A man named Anders Celsius spent more time thinking of a more useful, practical scale. A Celsius thermometer can be reconstructed from scratch with a minimum of equipment. Take an ungraduated (no markings) thermometer and place it in ice water. Mark the level of the thermometer fluid and label it with a zero. Then place the thermometer into boiling water and label the new level as "100." Finally, divide the space between the two marks with 100 even levels. Because there are 100 even levels between these two standard temperatures, the system developed by Anders Celsius is also referred to as the Centigrade ("100 levels") scale. The Celsius scale is the metric standard and is used by the science community and the majority of the world.

In the Celsius (Centigrade) scale:

- Water freezes at 0°C.
- Water boils at 100°C.

Kelvin

Lord Kelvin liked the simple elegance of the Celsius scale, but he took it one step further. Kelvin felt that it was a misrepresentation to have a zero on a scale if temperatures can get below zero. The casual observer might misinterpret a temperature of –10°C as having "negative energy" or "cold energy." In fact, there is (in Science) no such thing as "cold." Coldness is simply a lack of heat.

Ice doesn't add cold to something. It sucks the heat into it and then uses it for melting.

Realizing this, Kelvin designed a scale where the zero represented no kinetic energy at all. The zero on Kelvin's scale would represent the absolute coldest that a material could get—or **absolute zero**. Kelvin calculated at what Celsius temperature this would happen, assigned zero to it, and then used the "size" of the Celsius degrees to put levels on his thermometer. Thus, the freezing and boiling points of water are still separated by 100 levels.

Kelvin refers to an absolute change rather than "degrees" of change, so there is no degree symbol.

In the Kelvin scale:

- Absolute zero is –273° on the Celsius scale.
- Water freezes at 273 K. (There's no degree mark in Kelvin.)
- Water boils at 373 K.

Temperature Conversions

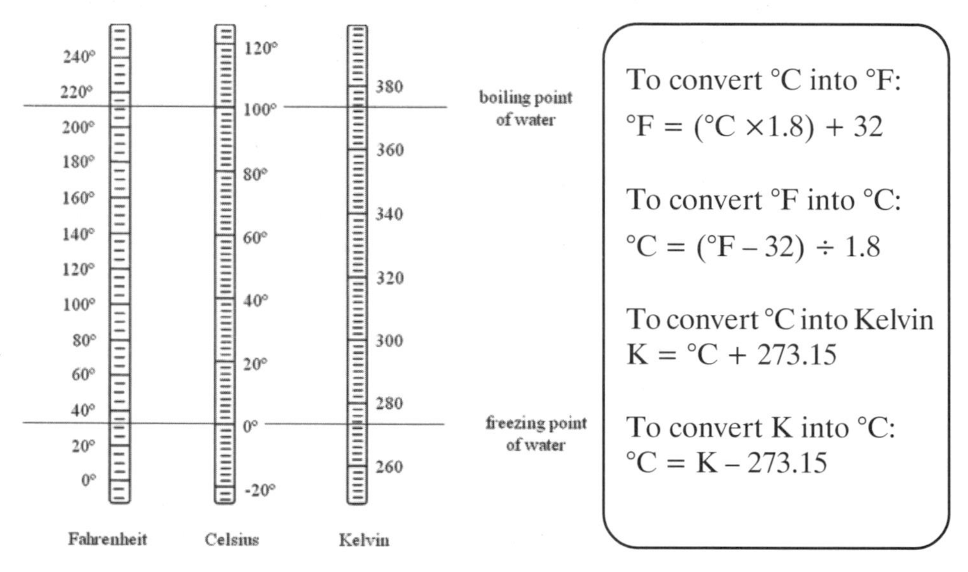

Figure 7.5 Temperature Scales

Lesson 7–2 Review

Exercise A

Convert the following from Fahrenheit to Celsius.

Fahrenheit	Celsius	Fahrenheit	Celsius
1. 32°F	________	4. 140°F	________
2. 98°F	________	5. 20°F	________
3. 212°F	________		

Exercise B

Convert the following from Celsius to Fahrenheit.

Celsius	Fahrenheit	Celsius	Fahrenheit
6. 80°C	________	9. 38°C	________
7. 20°C	________	10. 32°C	________
8. 12°C	________		

Exercise C

Convert the following from Celsius to Kelvin.

Celsius	Kelvin	Celsius	Kelvin
11. 0°C	_______	14. –20°C	_______
12. 100°C	_______	15. 75°C	_______
13. 50°C	_______		

Multiple Choice

16. To heat 1 g of water by 1°C, it will require how much energy?
    a) 1 calorie c) 50 watts
    b) 1,000 calories d) 1.0 watts

17. Specific heat describes an object's
    a) temperature. c) ability to heat up.
    b) temperature in Kelvin. d) average kinetic energy.

18. During a phase change, water's temperature
    a) stays the same. c) suddenly drops.
    b) shoots up quickly. d) drops gradually.

19. Heat gets from the Sun to the Earth by the process of
    a) subduction. b) conduction. c) convection. d) radiation.

20. Once the heat gets from the Sun to the Earth, most of the heat transfer around the planet is done by
    a) subduction. b) conduction. c) convection. d) radiation.

Lesson 7–3: Atmospheric Pressure

Air Pressure

Air pressure is the weight of the air above you. If you drew a 1-inch square on a piece of paper, try to imagine a column of air sitting on top of that square extending up to the top of the atmosphere. The weight of the air in that column would be 14.7 pounds under normal conditions. Now imagine the victim of a horrible prank—a person who woke up with a hangover only to see his entire body covered with a tattooed grid of 1-inch squares. On that person's body, there would be a few thousand square inches of surface area—each one with 14.7 pounds pushing in on it.

What about the squares that face downward towards the ground? Air is a fluid, which means that it can flow around corners. Therefore, the weight of the air is distributed all around you in every direction.

Why are we not crushed by the air pressure pushing all around us? We have the same amount of air pressure inside us trying to push out as is outside pushing in. There is a "push-of-war" happening on our skin. Your ears "pop" whenever one side pushes a little more than the other.

The device used to measure air pressure—also called barometric pressure—is the barometer. There are a few different ways to express air pressure, but they all represent the weight of the air. "Inches of mercury" or simply "inches" refers to how many inches of mercury air pressure will push up a vacuum tube. "Millibars" and "atmospheres" refer to "normal air pressure." Normal air pressure at sea level is one atmosphere worth of air and therefore called "one atmosphere." The millibar system assigns "one bar" (for *bar*ometric pressure) as normal air pressure. Air pressure changes occurring in weather are relatively small. Therefore, one bar is divided into 1,000 units called millibars.

Station Model Code

On a station model (see Figure 7.6), air pressure is written at the top left corner. It is expressed in millibars but to save space, it is written in code. To convert from millibars (mB) to "station model code" take the last three numbers from the barometric pressure in millibars. For example 998.2 mB turns into 982.

To convert from "station model code" to millibars, do the following:

- If the value is "500" or higher, put a 9 in front of it and add the decimal for the last digit.
- If the value is lower than "500," put a 10 in front of it and add the decimal for the last digit.

The more air that is above you the more pressure you will feel. This explains why there is less pressure on the tops of mountains. As you ascend, you move above some of the atmosphere. In fact, the densest part of the atmosphere is at the surface so air pressure falls off quickly as you go up.

However, just by staying in place in your home, the air pressure will change due to weather conditions. Waves high in the atmosphere will cause the air above you to "pile up" in some spots and "get thinner" in others. The piled-up portions of the atmosphere will push down on you a little harder and we will feel higher pressure. Because nature tries to balance everything out, the piled areas of the atmosphere will sink to try to even out. The thinner, lower-pressured sections of the atmosphere will rise up and we will feel lower pressure. These slight changes in pressure are usually detected only by barometers, but occasionally people with sensitive injuries can feel an approaching weather system.

Whether the air is rising or sinking will have significant effects on the weather. In a nutshell, rising air makes clouds and produces "weather"; sinking air destroys clouds and makes clear skies. As always, there are exceptions, but understanding the relationship between air pressure and its effects gives one tremendous insight into the workings of weather.

- Warm air is less dense. It rises, forming clouds and causing lower pressure.
- Cold air is denser and sinks, creating higher pressure.

Summary of the Station Model

The station model is a representation of the weather conditions at a particular spot on a map. It is essentially a circle with the information jammed in and around it. To save space, much of the data (cloud cover, wind speed, pressure) is put into code.

Following is a simplified station model with the key data points illustrated.

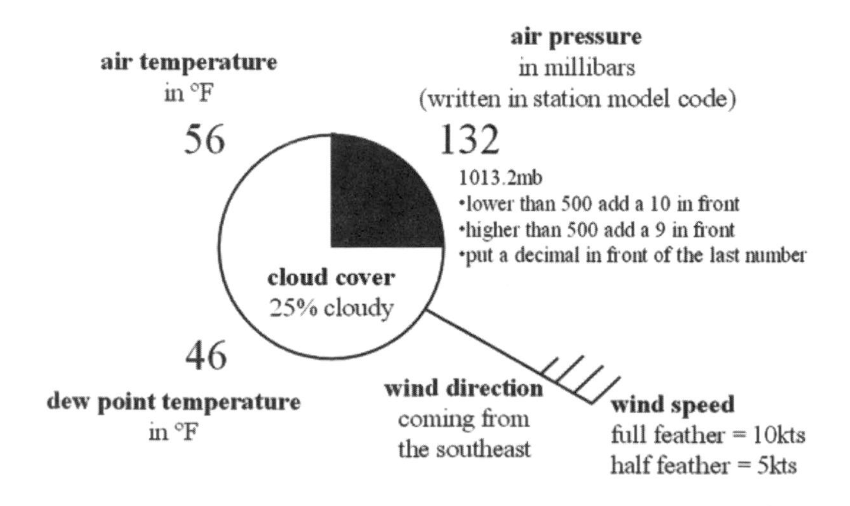

Figure 7.6. Station model

Lesson 7–3 Review

Exercise A

Complete the following chart by converting the millibars into station model code.

	Millibars	"Station Model"		Millibars	"Station Model"
1.	1028.0		6.	1000.1	
2.	1008.4		7.	1008.2	
3.	992.2		8.	987.1	
4.	976.6		9.	988.8	
5.	994.8		10.	1022.2	

Exercise B

Complete the following chart by converting the station model code into millibars.

	Millibars	"Station Model"		Millibars	"Station Model"
11.		281	16.		888
12.		206	17.		165
13.		080	18.		768
14.		168	19.		000
15.		800	20.		987

21. What is the relationship between air pressure and weather conditions?

22. Isolines on a weather map connecting points of equal pressure are called

    a) isotherms. b) isobars. c) isosceles. d) contour lines.

23. Regions where isobars increase in value as the center is approached are _________ areas.

    a) high pressure
    b) low pressure
    c) stormy
    d) cloudy

24. The air pressure on the top of a mountain will be ________ the bottom of the mountain.
    a) higher than
    b) equal to
    c) less than
    d) The is not enough information to answer.

Lesson 7–4: Atmospheric Moisture

Water enters the atmosphere by evaporation. To understand what is involved with evaporation we need to revisit what temperature means. Temperature is the *average* kinetic energy of a material. In other words, some of the molecules of water are hotter and some are cooler. In a cup of water, there will be a few molecules of water that are moving so fast that they have enough energy to turn into the gas phase. If these molecules are at the surface, they will break free and escape into the atmosphere, taking its extra heat with it. **Evaporation** has just taken place. This cools the water that is left behind, which is why we feel cool when water evaporates off of our skin. The warmer the water is to begin with, the more molecules there will be that have enough energy to evaporate.

If you look at the molecules of water vapor in the air with the same perspective, there will be some molecules that no longer have enough energy to stay in the gaseous phase. These molecules will condense on the nearest surface and produce a tiny droplet of liquid water (condensation).

There are two ways that water can enter the atmosphere: evaporation and sublimation (boiling is "fast evaporation"). **Sublimation** happens when molecules of a solid go directly into the gas phase without first becoming a liquid. Once again, this has to do with the "average kinetic energy." In an ice cube, most of the molecules of water will have very little movement. However, there will be a few that are very energetic—energetic enough to become a gas. These molecules will escape the ice and become atmospheric moisture. Ice cubes in your freezer will sublime. This is why old ice cubes often look as if someone only filled the tray half-way. The moisture escaped into the atmosphere in your freezer and you see a cloud of moisture spill out when you open the door.

Humidity in the air is a constant balance between evaporation and condensation. The primary factors are how much energy and water are available. In a warm environment, there will be more evaporation. Condensation will still happen, but at a lesser rate.

There are two types of humidity that are used to describe the moisture content of the atmosphere: absolute humidity and relative humidity. **Absolute humidity** is a measure of the mass of water in a volume of air. Absolute humidity is not used very much in meteorology, so relative humidity is the one we will concentrate on here.

The amount of water that the air can hold is called its **capacity**. Capacity depends on the temperature. Because the highest-energy molecules are the ones that evaporate and get into the air, warmer air can hold more moisture than cooler air. In other words, warm air has a higher capacity of moisture than cold air.

Relative Humidity

Relative humidity tells us how close the air is to being "full" of moisture and has more practical applications than absolute humidity. **Relative humidity** is a percentage of how much moisture is currently in the air as compared to the capacity of the air at that temperature. It is actually inaccurate to say that the air is "full" of moisture—air that is full of moisture is called a lake. The air reaches its capacity when the rate of evaporation and condensation equal each other. At this point the air is **saturated**.

To measure relative humidity, you need an instrument called a sling psychrometer. The **psychrometer** is simply two thermometers: a regular one (called the "dry bulb" thermometer) and one with a moistened cloth over the bulb at the bottom (called the "wet bulb" thermometer). In principle, here's how it works: When water evaporates off of the wet cloth it also takes away some of the heat in the thermometer. This makes the thermometer colder than the air and "depresses" the temperature. The amount of the wet bulb depression directly depends on how much moisture can evaporate into the air. If the temperature dropped by a little, there wasn't much evaporation and the air is close to saturation. If the temperature dropped by several degrees, the air is dry.

Dew Point

Dew point and relative humidity are very closely related. They both give an indication of how much water is in the atmosphere. **Dew point** is the temperature that you will need to chill the air, under current conditions, in order for dew (condensation) to form. If you drink a glass of soda, the outside of the glass will "sweat" once the temperature of the outside of the glass reaches the dew point. If it is currently dry at ground level on a 70°F day and the dew point is 50°F, clouds will be forming at an altitude where the ground-level air has cooled to 50°F.

To Use a Relative Humidity Chart

The important thing to do when using this chart is to read every word on the chart before you start work. The big mistake that people make is to read the numbers at the top of the chart and think that they are the wet-bulb or the dew point values. At the top of the chart, it says "difference between wet-bulb and dry-bulb." You must read and understand that this means that you have to subtract. Once you get past that trap, the rest is easy.

Relative Humidity (%)

Dry-Bulb Temperature (°C)	Difference Between Wet-Bulb and Dry-Bulb Temperatures (C°)															
	0	1	2	3	4	5	6	7	8	9	10	11	12	13	14	15
−20	100	28														
−18	100	40														
−16	100	48														
−14	100	55	11													
−12	100	61	23													
−10	100	66	33													
−8	100	71	41	13												
−6	100	73	48	20												
−4	100	77	54	32	11											
−2	100	79	58	37	20	1										
0	100	81	63	45	28	11										
2	100	83	67	51	36	20	6									
4	100	85	70	56	42	27	14									
6	100	86	72	59	46	35	22	10								
8	100	87	74	62	51	39	28	17	6							
10	100	88	76	65	54	43	33	24	13	4						
12	100	88	78	67	57	48	38	28	19	10	2					
14	100	89	79	69	60	50	41	33	25	16	8	1				
16	100	90	80	71	62	54	45	37	29	21	14	7	1			
18	100	91	81	72	64	56	48	40	33	26	19	12	6			
20	100	91	82	74	66	58	51	44	36	30	23	17	11	5		
22	100	92	83	75	68	60	53	46	40	33	27	21	15	10	4	
24	100	92	84	76	69	62	55	49	42	36	30	25	20	14	9	4
26	100	92	85	77	70	64	57	51	45	39	34	28	23	18	13	9
28	100	93	86	78	71	65	59	53	47	42	36	31	26	21	17	12
30	100	93	86	79	72	66	61	55	49	44	39	34	29	25	20	16

Figure 7.7

Example:

Air Temperature = 18°C Wet-bulb = 14°C

- ▷ Take the current air temperature, that's the dry-bulb, and slide down the side of the chart until you see that temperature. If it is an odd number, just go down to the next printed value for now.

- ▷ Slide across to the difference between the two thermometers: 4°C.
- ▷ The box you are in is the relative humidity: 64%
- ▷ If you had an odd number for the temperature, just average this number and the one in the box directly above.

To Use a Dew Point Temperature Chart

The dew point temperature chart works *exactly* the same, so there's no new reading here. Very often, problems will ask you to solve for the relative humidity and dew point temperature for a set of temperatures. The beauty of this is that you only have to work the charts once, and both answers will be in the same square on each chart.

Dewpoint Temperatures (°C)

Dry-Bulb Temperature (°C)	Difference Between Wet-Bulb and Dry-Bulb Temperatures (C°)															
	0	1	2	3	4	5	6	7	8	9	10	11	12	13	14	15
−20	−20	−33														
−18	−18	−28														
−16	−16	−24														
−14	−14	−21	−36													
−12	−12	−18	−28													
−10	−10	−14	−22													
−8	−8	−12	−18	−29												
−6	−6	−10	−14	−22												
−4	−4	−7	−12	−17	−29											
−2	−2	−5	−8	−13	−20											
0	0	−3	−6	−9	−15	−24										
2	2	−1	−3	−6	−11	−17										
4	4	1	−1	−4	−7	−11	−19									
6	6	4	1	−1	−4	−7	−13	−21								
8	8	6	3	1	−2	−5	−9	−14								
10	10	8	6	4	1	−2	−5	−9	−14	−28						
12	12	10	8	6	4	1	−2	−5	−9	−16						
14	14	12	11	9	6	4	1	−2	−5	−10	−17					
16	16	14	13	11	9	7	4	1	−1	−6	−10	−17				
18	18	16	15	13	11	9	7	4	2	−2	−5	−10	−19			
20	20	19	17	15	14	12	10	7	4	2	−2	−5	−10	−19		
22	22	21	19	17	16	14	12	10	8	5	3	−1	−5	−10	−19	
24	24	23	21	20	18	16	14	12	10	8	6	2	−1	−5	−10	−18
26	26	25	23	22	20	18	17	15	13	11	9	6	3	0	−4	−9
28	28	27	25	24	22	21	19	17	16	14	11	9	7	4	1	−3
30	30	29	27	26	24	23	21	19	18	16	14	12	10	8	5	1

Figure 7.8

Example (same conditions as the relative humidity example, but now we're finding the dew point temperature):

Air Temperature = 18°C   Wet-bulb = 14°C

- ▷ Take the current air temperature, that's the dry-bulb, and slide down the side of the chart until you see that temperature. If it is an odd number, just go down to the next printed value for now.

- Slide across to the difference between the two thermometers: 4°C.
- The box you are in is the dew point temperature: 11°C.
- If you had an odd number for the temperature, just average this number and the one in the box directly above.

In order for condensation to occur, water needs a surface on which to condense. At ground level, you can see this as dew on blades of grass or on the hoods of cars. However, in the air, you rarely find blades of grass or car hoods, yet condensation still happens in large amounts. The surfaces that water condenses onto are usually small particles of dust, smoke, or salt (from sea spray). When acting as a surface for condensation, these particles are called **condensation nuclei**. Normally, the atmosphere has enough condensation nuclei to allow cloud formation. However, artificial nuclei can be added by "seeding" the clouds with smoke, which may stimulate more condensation. After a rain, the atmosphere is usually cleaner and clearer because dust and other pollutants have been stripped from the air.

Dry Adiabatic Lapse Rate

As you climb into the atmosphere, the temperature gets colder; this effect is called the **lapse rate**. The lapse rate for dry air is the **adiabatic lapse rate**, which is 5.5°F colder for every 1,000 feet of altitude. The dew point also gets lower as you climb into the air, but not as quickly. This is called the **dew point lapse rate** and it equals 1°F for every 1,000 feet you go up. Clouds will form where the two values (the dew point temperature and air temperature) are equal.

As air rises along with its moisture, the two temperatures get closer and the air gets more humid (that is relative humidity gets closer to 100%). If the air rises enough, it will reach a point where the air temperature and dew point are equal. Water will start to condense and will continue to do so as the air continues to rise. The altitude where the condensation began is the **cloud base height,** and this explains why clouds often have a flat bottom and billow upwards.

Lesson 7–4 Review

Exercise A

Complete the following chart using the relative humidity and dew point temperature charts.

	Dry Bulb	Wet Bulb	Relative Humidity	Dew Point Temp.
1.	12°C	7°C		
2.	22°C	20°C		
3.	18°C	12°C		
4.	6°C	5°C		
5.	22°C			21°C
6.	20°C		66%	
7.	19°C	19°C		
8.	17°C	13°C		
9.	26°C	24°C		
10.	16°C	12°C		

Exercise B

For each of the following, write either **evaporation** or **condensation**.

11. The process by which a substance changes from a liquid to a gas is called ________________.
12. During ________________ more molecules break free from a liquid than join it.
13. When more molecules join a liquid than leave, ________________ takes place.
14. When molecules of water vapor collide and stick together in the air, ________________ occurs.

From the following list, choose the term that best completes each sentence.

evaporation	melting	kinetic energy
sublimation	freezing	condensation

15. The higher the temperature of something, the greater its ________.
16. The change in state from gas to liquid is called ________________.
17. Liquid changing to gas only at the surface is called ______________.
18. The change in state from solid to liquid is called _______________.
19. The change in state from liquid to solid is called _______________.
20. In _________________ particles pass directly from solid to gas.

Lesson 7–5: Wind

Wind is the horizontal movement of air across the surface. Air is forced across the surface between areas where the air is rising and sinking. In other words, wind is created between high pressure and low pressure areas. If there is a small difference, or gradient, in pressure between the two pressure systems, then the wind will be gentle. If the gradient is greater, then the wind will be stronger.

Winds are named for the direction from which they come. A wind coming from the north will be a "north wind." This makes sense because the wind brings the weather with it. A north wind will bring the northern weather with it and will usually be a cold wind. Likewise, a south wind will be warm. On a weather map, wind direction will be shown with an arrow. The arrow points in the same direction that wind blows, but the head of the arrow is usually buried in the circle that marks the weather station's location. This leaves the feathered tail of the arrow shaft exposed. The feathered tail points to the direction that the wind came from, which gives us its name.

	5kts
	10kts
	20kts
	35kts
	50kts
	75kts
	115kts

Figure 7.9. Wind speed

Wind speed is measured in "knots." A **knot** is a nautical mile per hour and is slightly faster than a "regular" mile per hour. On a weather map, wind speed is shown with the feathers on the arrow shaft (see chart). Each feather represents 10 knots. A half feather is worth 5 kts. To help tell the difference between a full 10 kt feather and a smaller 5 kt half-feather, if there is only a single half-feather it is drawn in a little from the end of the arrow shaft. Each 50 knots gets a small flag.

Winds are also ranked in intensity using the Beaufort scale. The **Beaufort scale** rates the wind speed from 0 to 12 based on simple observations.

Beaufort Number	**Wind Speed** mi/hr(km/hr)	**Description**
0 Calm	less than 1 (less than 1.6)	Smoke will rise vertically.
1 Light Air	1–3 (1.6–4.8)	Rising smoke drifts, weather vane is inactive.
2 Light Breeze	4–7 (6.4–11.3)	Leaves rustle, you can feel wind on your face, weather vane is inactive.
3 Gentle Breeze	8–12 (12.9–19.3)	Leaves and twigs move around, light weight flags extend.
4 Moderate Breeze	13–18 (20.9–29.0)	Moves thin branches, raises dust and paper.
5 Fresh Breeze	19–24 (30.6–38.6)	Trees sway.
6 Strong Breeze	25–31 (40.2–50.0)	Large tree branches move, open wires (such as telephone wires) begin to "whistle," umbrellas are difficult to keep under control.
7 Moderate Gale	32–38 (51.5–61.2)	Large trees begin to sway, noticeably difficult to walk.
8 Fresh Gale	39–46 (62.8–74.0)	Twigs and small branches are broken from trees, walking into the wind is very difficult.
9 Strong Gale	47–54 (75.6–86.9)	Slight damage occurs to buildings, shingles are blown off of roofs.
10 Whole Gale	55–63 (88.5–101.4)	Large trees are uprooted, building damage is considerable.
11 Storm	64–72 (103.0–115.9)	Extensive widespread damage. These typically occur only at sea, and rarely inland.
12 Hurricane	>73 (>115.9)	Extreme destruction.

The Coriolis Effect

The **Coriolis Effect** deflects things (fluid things, such as air and water) to their right in the Northern Hemisphere. It is caused by the rotation of the Earth and the spherical shape of the planet.

Here's an analogy that may help to visualize how the Coriolis Effect works. You have a snowball fight with Santa Claus. You are at the equator and Santa, of course, is at the North Pole, and both of you are equipped with a super snowball cannon. You line up the crosshairs of your snowball cannon directly on Santa's nose and fire. Much to your chagrin, your snowball landed to Santa's right. This was caused by the Earth's rapid rotation (about 1,000 mi/hr) at the equator. You, and your snowball, are moving at close to 1,000 mi/hr to the right, while Santa, at the North Pole is standing still rotating in place. If your snowball takes an hour to reach the North Pole, your shot will land roughly 1,000 miles to the right. In addition, the path that the snowball took over the surface will be a gentle curve deflected to the right of its travel direction.

Santa now sends a volley your way. He waits until you are directly in his cannon's sights and fires. Santa, and his snowball, are standing still while you are on the equator and sliding to the right at 1,000 mi/hr. If it takes Santa's snowball an hour to reach the equator, you would have traveled an additional 1,000 miles to the right and the snowball lands harmlessly behind you. From Santa's point of view, his snowball curved to *its* right to land behind you.

This effect causes large-scale planetary winds to get deflected to their right in the Northern Hemisphere. The opposite is true in the Southern Hemisphere.

The Coriolis Effect is strongest in the middle latitudes (where the United States is located), which explains why the majority of spinning storms (tornadoes and hurricanes) occur in this region. The effect is weakest at the equator. As a result, this region is free of organized storm systems.

What's the deal with the Coriolis Effect and spinning toilet bowls: Do toilets in Australia spin the opposite way?

The Coriolis Effect is a large-scale, slow-working effect. A toilet is too small and works too quickly for the effect to happen. In addition, the water in the bowl gets forced in one direction or the other by the shape of its drain and the direction that the fresh water is injected.

Pressure Systems

A low pressure system is caused by rising air. As the air rises it pushes less on us and our weather instruments and is recorded as lower pressure. Because it is a rising air system, a low pressure center will usually be made of warmer air than its surroundings. When the air rises, it sucks in more surrounding air to replace the air that has risen. The air around a low pressure center gets pulled in, or **converges**, and then rises. As the air gets sucked in it is deflected and begins to spin counterclockwise.

The inward and counterclockwise airflow of a low pressure system is called a **cyclone**. The word cyclone is often misused or misunderstood to refer to a tornado or a hurricane. In fact, any low pressure system, of which tornadoes and hurricanes are only two examples, is a cyclone. The term only refers to the characteristic air flow: inward and counterclockwise.

Low Pressure Sucks!

Low pressure creates a sucking effect.

Also, the rising air of low pressure creates clouds and foul weather. Therefore, low pressure sucks!

A high pressure area is created by sinking air, which is cooler than the surrounding air. The sinking air pushes down on us and our instruments a little harder and we record higher pressure. As the cold air sinks, it gets compressed by the increased pressure near the ground. The compression warms the air slightly. Warmer air can hold more moisture and causes the relative humidity to drop. The air "gets drier" and clouds do not form.

Note: Colder, sinking air is high pressure—no clouds.

When high pressure air reaches the ground, it spreads out in all directions just as if you poured water onto a table top. As the air spreads out, or **diverges**, it gets deflected by the Coriolis Effect. As a result, a high pressure system will be divergent and the air will flow clockwise out from the center.

The airflow around a high pressure center, outward and clockwise, is called an **anticyclone**. Because a high pressure center is descending air and devoid of clouds, the anticyclone is not associated with any particular storm.

The Right-Hand Rule

One way that helps keep track of the all directions in which the air around pressure systems moves is to use a trick called the "right-hand rule" which is borrowed from physics class. The thumb on your hand represents the direction (up or down) that the air moves—for example, "up" for a low pressure system. When you make a "thumbs-up" sign with your right hand, the fingers will wrap around your palm counterclockwise. Combine this with the easy-to-remember "low pressure sucks," and you have it all. In a low pressure system, warm air rises, gets sucked inward, and spins counterclockwise.

It works in the same fashion for high pressure. In a high pressure system, the air (your right thumb) sinks. The air blows outward and your fingers now wrap around your palm clockwise.

A combination of the Coriolis Effect and the different temperatures around Earth creates wind belts that wrap around the planet. The wind belts give us the prevailing winds. A **prevailing wind** is the "typical" direction from which winds come for a particular region of the world. In the United States, most areas have a prevailing southeasterly wind.

The winds that carried Columbus across the Atlantic blew from the east, and later Europeans used the winds to carry them across the ocean for trade. These winds are called the trade winds.

Prevailing winds are very important for predicting the weather. If the prevailing winds are from the west, a good indication of what the weather will be to look to the west. If it is a storm on the horizon or a large system moving across the country all you need to do is see if it is to the west.

Lesson 7–5 Review

1. What is the relationship between pressure gradient and wind speed?
   a) inverse
   b) direct
   c) cyclic
   d) They are independent of each other.

2. In the Northern Hemisphere, the Coriolis Effect will cause the winds around a low pressure system to flow
   a) inward and clockwise. c) inward and counterclockwise.
   b) outward and clockwise. d) outward and counterclockwise.
3. Winds are named after the direction they are blowing
   a) toward. b) from. c) into.
4. Wind speed on a weather map is indicated by
   a) a long dotted line shaped like an arrow.
   b) a number written in the lower left corner of the station model.
   c) "feathers" drawn on the shaft of an arrow.
   d) "arrow shafts" sticking out of the station model.
5. Which statement describes the relationship between air pressure and weather conditions?
   a) Low pressure sinks downward and blows out, clearing out any clouds.
   b) Winds blow from low pressure to high pressure, pushing the weather along.
   c) Low pressure sucks the air inward and up, creating clouds and bad weather.
   d) High pressure sucks the air inward and up, creating clouds and bad weather.

Lesson 7–6: Air Masses and Fronts

An **air mass** is a large region of air that has the same characteristics of temperature and humidity. Think of it as a large bubble of air. For example, the air in a tropical air mass will usually be hot and humid.

An air mass gets it characteristics from the region from which it comes—called its **source region**. As air moves across the surface of the Earth it picks up some of the characteristics of the surface. If it is over the tropical ocean it will get warmer and more moist; if it is over central Canada, it will be cool and dry. Inside an air mass, the weather conditions will be somewhat stable. The temperature will stay consistent as will the humidity levels.

When describing an air mass, four terms (which refer to the source regions) are used: two to describe the humidity within the air mass, and the other two to describe the temperature. With respect to the humidity the two terms are *continental* and *marine.* **Continental** describes a dry air mass that took on the characteristics of the land. A **marine** air mass was formed over the ocean and is therefore moist.

To express the types of temperatures air masses have, the two terms are *tropical* and *polar*. A **tropical** air mass is warm and usually comes from southern regions; **polar** is a cooler air mass and migrated from the north. Sometimes, the term **arctic** is used to describe an extremely cold air mass.

By combining these terms we can describe weather conditions within an air mass.

Air Mass Name	Characteristics	Map Symbol
continental tropical	dry and warm	cT
continental polar	dry and cold	cP
continental arctic	dry and very cold	cA
maritime tropical	moist and warm	mT
maritime polar	moist and cold	mP

Fronts

As previously mentioned, the air inside an air mass is fairly stable. A mass is filled by air that has similar properties. As a result, no sections of air are rising or sinking rapidly which is the cause of turbulent weather. However, on the edges of the air mass, there are usually contrasts in conditions: warm air meets cooler, moist air meets dry. When the two air masses collide, the differences in densities (warm is lighter than cool and moist is lighter than dry) cause the air to rise and fall, which creates clouds and rain. In maps edges of air masses are often marked with lines or regions of clouds.

The passage of one air mass and the entrance of the next is usually heralded by a period of rain (or at least more cloud cover) followed by a change in temperature. The exact place where the two air masses meet is called a front. A **front** can also be thought of as the front edge of an incoming air mass. Fronts get their names from the type of air that is coming in. For example, if you were in warm air, then it rained and got colder, you just experienced the passage of a cold front. You were in a warm air mass and now you're in a cold air mass.

The type of weather associated with a front comes from the motion of the air as the two air masses collide.

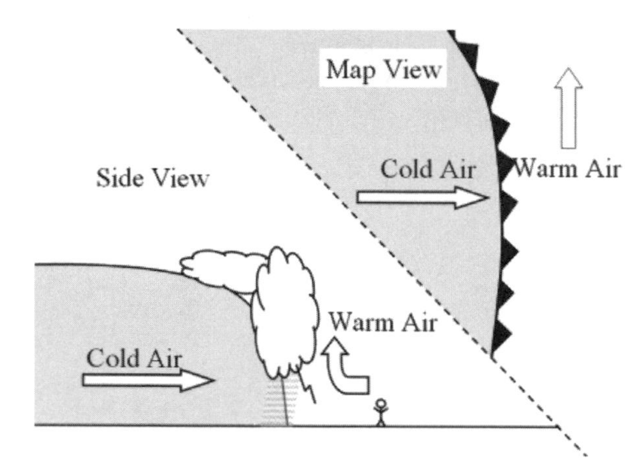

Figure 7.10. Cold Front

A **cold front** is the leading edge of a cold air mass. The incoming cold air is denser than the warmer air that is being pushed out of the way. As a result, once a cold front gets moving, it will move faster than a warm front. The front, as seen from the side, takes on the shape of the front edge of cold, thick syrup poured onto a table. The warm, light air in its way will be forced to rise straight upward very quickly. The passage of a cold front will often bring a sudden, short period of heavy rain. If conditions are right, the vertical cloud development will result in the formation of thunderstorms.

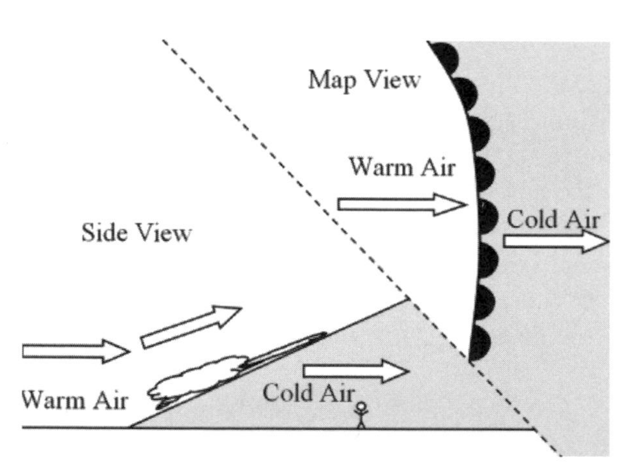

Figure 7.11. Warm front

A **warm front** is the leading edge of an incoming warm air mass. When the warm air pushes on the dense cold air in front of it, the warm will be deflected upward and has a harder time pushing the cold dense air out of the way. Warm fronts move slower than cold fronts. The place where the two air masses meet on the ground is the exact spot where the front is drawn on a map, but the first effects of the front stretch well in front of the front. The first indication that a warm front is approaching is the appearance of cirrus clouds (high, thin, whispy clouds). As the front gets closer, the clouds will lower and thicken. Several hours before the front reaches your location, the clouds thicken enough to support rain and it will rain gently until the front reaches you.

On a weather map, the symbol for the cold front is a line connecting triangles. Think of the triangles as icicles. The warm front is a line of half-circles—like melted icicles.

An **occluded front** is a combination of a warm and cold front. As discussed earlier, cold fronts usually travel faster than warm fronts. Sometimes the cold front will catch and override the warm front. When the two fronts pile up on each other, it produces the combined effects of both fronts. First, the effects of the warm front are experienced by observers: high, thin, wispy clouds; then clouds lower and thicken; finally a long period of gentle rain. After the effects of a warm front, the occluded front produces cold front characteristics: a short period of heavy rain with the possibility of thunder.

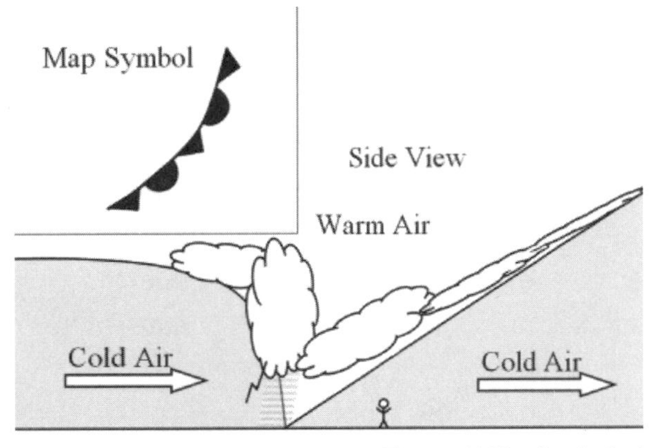

Figure 7.12. Occluded

Occluded fronts will often be formed near the center of a low pressure system. As the low spins counterclockwise, the two fronts, warm and cold, get "wrapped up" together.

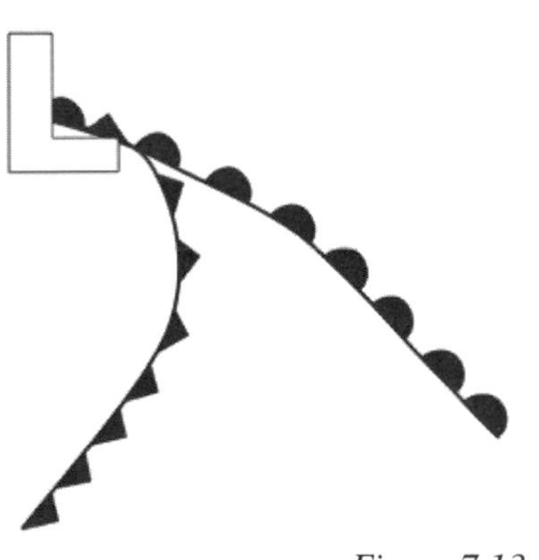

Figure 7.13

The final type of front is the stationary front. The **stationary front** marks the location where two air masses meet but neither one advances on the other. This could be caused by a literally stationary system. It can also be caused by the movement of the air masses that is parallel to the line dividing them—in effect, the front is sliding sideways.

When dealing with stationary fronts it tends to be very difficult to predict the weather associated with them. The front marks the location where cold and warm air meet so there will be some mixing and vertical motion (convection). If the front twitches one way or the other, it will produce the effects of a warm front or a cold front depending on which air mass advances slightly. Often, this type of system is referred to as "unstable."

Lesson 7–6 Review

1. When a warm front moves into an area,
   a) usually the skies are gray and there is often a drizzle.
   b) there is often heavy precipitation followed by clearing when the front has moved through.
   c) there is usually little change in the weather.
2. Maritime tropical (mT) air masses form
   a) over water closer to the equator.
   b) over land closer to the equator.
   c) over cold waters.
   d) over cold land.
3. In front of a cold front, the temperatures will be
   a) colder than they are behind the front.
   b) warmer than they are behind the front.
   c) the same temperature as those behind the front.
   d) either colder or warmer than the temperature behind the front.
4. As a warm front approaches, it will often rain
   a) before the front reaches your location.
   b) just as the front reaches your location.
   c) after the front passes your location.
   d) to the west of the front.
5. Thunderstorms are mostly associated with which type of front?
   a) cold front b) warm front c) stationary front

Lesson 7–7: Storms

Thunderstorms

A **thunderstorm** comes from a well developed cumulonimbus cloud also know as a **thunderhead**. These tall, billowing clouds are made from strong updrafts of warm, moist air and can reach in excess of 40,000 feet high. The special effects of these storms come from the intense turbulence within the cloud. Rapidly rising and falling air cause friction between the air currents. As with friction between a carpet and your socks, the result is the build-up of static electricity. When the electricity builds up enough strength it will jump across the air in an effort to find an opposite charge and neutralize. This is the stroke of lightning.

Basically, **lightning** is caused by the cloud's air currents separating positive and negative charges within the cloud. One side of the cloud, for example the bottom, will take on a positive charge while the top will have a negative charge. The tendency is for the two charges to come back together and neutralize. This is difficult because the air in between the two sides acts as an insulator. If the charge builds up enough, the electricity will jump through the air and neutralize; a bolt of lightning happens within the cloud. This is called *cloud-to-cloud lightning*, which is the most common type of lightning.

One thing to consider, however, is the height of the cloud. If the top of the cloud, with a negative charge, is at 40,000 feet while the bottom of the cloud, with a positive charge, is at 1,000 feet, the bottom of the cloud is much closer to the ground than the top of the cloud. In this case, the positive charges of the cloud bottom attract the negative charges in the ground. If the attraction builds enough, the charges will jump across the air and lightning will strike the ground.

Lightning is lazy! It searches for the easiest path to neutralize itself. When the charges reach a critical level, both the cloud and ground will send out invisible streamers towards each other. When the streamers meet, the electricity has found the path of least resistance and the electricity flows. Air is very difficult for electricity to flow through. It is much easier for electricity to flow through trees and buildings and people. If the lazy lightning can make part of its trip through any of these objects, it will usually do so. This is why high objects tend to get struck by lightning.

Is it safe to stand under a tree during a thunderstorm?

That depends. What you want to avoid is standing under any object that sticks up higher than the surrounding objects. If it is one solitary tree in a field, then don't stand under it. If you are in a forest, then each tree has just as much of a chance of getting struck as any other tree. Besides, if you run out of the forest into a field, you are now the highest object around!

Besides the obvious lightning aspect of thunderstorms, another hazard that can come from a thunderhead is **hail**. Because the cumulonimbus clouds are created by violent updrafts and downdrafts, drops of water can get caught in the currents. As the droplets travel upwards, they will pass the point where the temperature cools to below freezing and turn into a pellet of ice. When the ice drops, it will collect a layer of liquid water over the ice. If it gets caught in another updraft it will be pushed to the freezing heights of the cloud again. This cycle will continue to repeat until the hail stone gets too large for the updraft lift again and the stone falls.

Mid-Latitude Low

A mid-latitude low is a typical, run-of-the-mill storm system. Simply put, it is a low pressure system that has organized itself into a weather system.

It will often be comma-shaped: a warm front with a cold front extending from the low. As the storm develops, the cold front will catch up to the warm front and form an occluded front near the center. Depending on the storm track, you will experience the warm front then the cold front, or the conditions of an occluded front.

The most intense portion of the mid-latitude low is the northern part where the low is located. In this area, the winds wrap around from the northeast and give the storm one of its nicknames: a **nor'easter**. Sometimes, winter mid-latitude lows are called **Alberta Clippers** because they come out of Alberta, Canada, and move quickly like the old clipper sailing ships.

Hurricane

A **hurricane** is a large cyclonic, low pressure system that originates in tropical regions. The storms can be hundreds of miles across and carry winds registering at more than a hundred miles per hour. A hurricane is characterized by its round, whirlpool shape, inward and counterclockwise airflow, strong winds, and heavy rain. A hurricane will often have an "eye" or calm center with little or no clouds.

Most hurricanes that affect the United States originate off the west coast of Africa, the tropical Atlantic, or the Caribbean. The storms begin as strong thunderstorms or "**tropical depressions**" and continue to gain strength. The majority of a hurricane's energy comes from the warm, moist ocean water over which it forms. The low density of the rising warm moist air creates low pressure. To add a boost to the storm, extra heat is released into the storm from condensation. As the pressure drops, the storm's intensity increases.

As the storm develops, it changes from a tropical depression to a **tropical storm** when its sustained winds get faster than 38 miles per hour. If the storm continues to strengthen and its winds pass 73 miles per hour, it will officially become a hurricane.

The rating system for hurricanes is the **Saffir/Simpson Hurricane Scale**. It rates a hurricane based primarily by wind speed from a category I (weakest) to a category V (strongest).

Saffir/Simpson Hurricane Scale

Category	**Wind speed** (mi/hr)	**Storm Surge** (ft)
Tropical Depression	Less than 38	
Tropical Storm	39–73	
I	74–95	4–5
II	96–110	6–8
III	111–130	9–12
IV	131–155	13–18
V	Faster than 155	More than 18

Although a hurricane's winds can be very dangerous and damaging, the most destruction and deaths come from the fury of the water during the storm. The intense low pressure associated with the hurricane actually sucks the ocean water upwards a little, creating a storm surge. The **storm surge** is a "dome" of water with its level higher than the surrounding ocean. When the hurricane makes landfall, it will carry the storm surge onto land, flooding coastlines. If a storm surge is coupled with a high tide, the effects can be devastating.

A hurricane that forms over the Atlantic Ocean is called a *hurricane*. A hurricane that forms over the Pacific Ocean is called a *typhoon*. Both storms are exactly the same with one exception: The Pacific Ocean is larger, which gives the storms more of an opportunity to gain strength. Therefore typhoons have the potential to be stronger than hurricanes.

Tornado

A tornado is a different type of spinning storm. It is much more localized and affects a much smaller area than a hurricane. Tornadoes are born out of strong thunderstorms called super cells. A **tornado** is a compact rotating storm with very strong winds.

Tornadoes are ranked on the Fujita scale. The scale ranges from 1 to 5, with 5 being the most intense. Some meteorologists also classify an F0 and an F6. An F0 is simply a circular wind that is weaker than 74 mi/hr. As incredible as it sounds, an F6 is difficult to detect. This is because the evidence that researchers would need to see to classify a tornado as an F6 will get destroyed by the weaker F5 winds just outside the strongest part of the storm.

Fujita Scale Number	**Wind Speed mi/hr** (km/hr)
F1	74–112 (118–180)
F2	113–157 (181–251)
F3	158–206 (252–330)
F4	207–260 (331–417)
F5	>261 (>418)

Lesson 7–7 Review

1. What factor is responsible for the general direction in which low pressure centers move across the United States?
   a) The United States is in a belt of prevailing westerly winds.
   b) The United States is in the horse latitudes.
   c) The United States is in the tropics.
   d) The United States is surrounded by water.
2. What is the relationship between air pressure and wind velocity in a tropical cyclone?
   a) As air pressure decreases wind velocity decreases.
   b) As air pressure decreases wind velocity increases.
   c) As air pressure changes there is no change in wind velocity.
3. What is the source of a hurricane's energy?
   a) a land mass
   b) the atmosphere
   c) a warm body of water
   d) lightning

4. An example of a mid-latitude low would be a
   a) thunderstorm.
   b) hurricane.
   c) tornado.
   d) nor'easter.

5. Most people who die in a hurricane die because of
   a) being hit by the strong winds.
   b) getting hit by large objects thrown by the wind.
   c) drowning in the storm surge.
   d) getting struck by lightning.

Chapter Exam

Exercise A

Complete the following chart using the relative humidity and dew point temperature charts.

	Dry Bulb	Wet Bulb	Relative Humidity	Dew Point Temp.
1.	30°C	21°C		
2.	0°C	–2°C		
3.	25°C	21°C		
4.	11°C	6°C		
5.	26°C			15°C
6.	8°C		17%	
7.	20°C			14°C
8.	16°C			–1°C

Exercise B

Complete the following chart.

	Temp. °F	Dew Pt. Temp °F	Air Press. mb.	Sky Cover	Wind Speed	Wind Direct.
9. 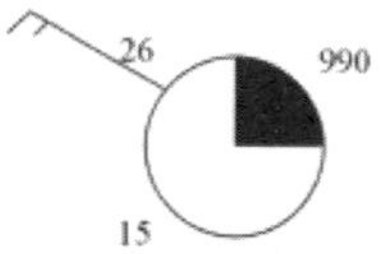 26 990 15						
10. 43 001 33 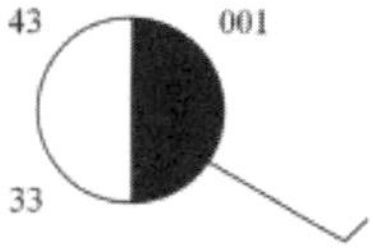						
11. 32 200 -8 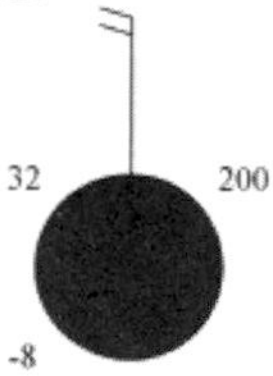						
12. 43 150 18 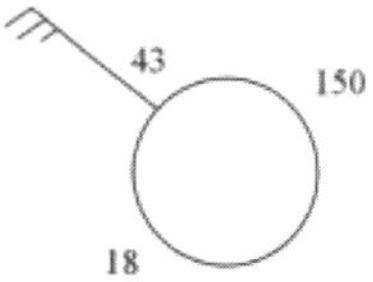						
13. 24 111 20 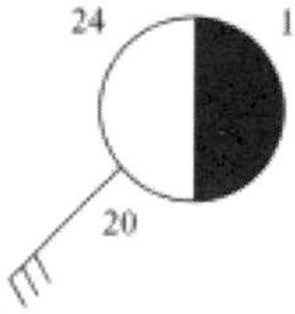						

Exercise C

For questions 14–22, refer to the Figure 7.19, which shows the hourly surface temperature, dew point, and relative humidity for a 24-hour period during the month of May. Note that all data are graphed using the same numbers on the left axis.

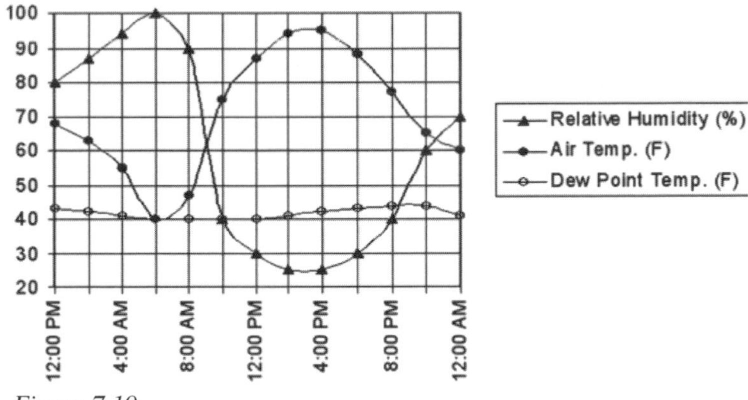

Figure 7.19

14. The lowest air temperature occurred at what time?
15. The highest relative humidity reading occurred at what time?
16. The highest air temperature occurred at about what time?
17. The lowest relative humidity occurred at about what time?
18. According to the graph, what happens to the relative humidity as the air temperature increases?
19. At what time(s) did the air temperature equal the dew point?
20. What was the relative humidity (%) when the air temperature equaled the dew point?
21. Condensation (water coming out of the air) is most likely to occur at approximately what time?
22. At approximately what time was the rate of evaporation highest?

Exercise D

Draw the station models for the following.

23. temp 30°F
    DPT 29°F
    Wind 10 kt from NW
    pressure 1012.0 mb
24. temp 54°F
    DPT 41°F
    Wind 15 kt from E
    pressure 1013.2 mb
    clouds 25%
25. temp 78°F
    DPT 78°F
    wind 0 kt
    pressure 986.4 mb
    overcast
26. temp 15°F
    DPT 15°F
    Wind 35 kt from NE
    Overcast
    pressure 1006.5 mb
27. temp 22°F
    DPT 18°F
    pressure 1021.0 mb

Exercise E

Decide whether each statement is true or false. In the space provided, write **T** or **F**. If the statement is false, correct the underlined word or words to make the statement true.

28. ___ Warm air can hold more water vapor than cold air.
29. ___ Cold fronts are usually associated with long periods of gentle rain.

30. ___ When the air temperature is very close to the dew point temperature the air is very <u>dry</u>.

31. ___ When saturated air is <u>cooled</u>, some of the water vapor will condense to form clouds or precipitation.

32. ___ A description of weather conditions at a particular weather station is called a <u>weather map</u>.

Answer Key

Answers Explained Lesson 7–1

1. **A** is **correct** because the troposphere contains the vast majority of water in the atmosphere.
2. **A** is incorrect because a watch, in meteorology, is a cautioning that the conditions are right for severe weather.

   B is **correct**.

   C is incorrect because relative humidity is a measure of how much moisture is in the air.

   D is incorrect.
3. **A** is incorrect because, even though oxygen is the most important gas for us, it is only a small part of the atmosphere.

   B is incorrect because carbon dioxide is less than 1% of the air.

   C is **correct** because nitrogen is 78% of the atmosphere.

   D is incorrect because methane is a very small fraction of the atmosphere.
4. **C** is **correct** because carbon dioxide absorbs and traps heat in our atmosphere.

Answers Explained Lesson 7–2

Exercise A

	Fahrenheit	**Celsius**
1.	32°F	**0°C**
2.	98°F	**37°C**
3.	212°F	**100°C**
4.	140°F	**60°C**
5.	20°F	**7°C**

Exercise B

	Celsius	Fahrenheit
6.	80°C	**176°F**
7.	20°C	**68°F**
8.	12°C	**54°F**
9.	38°C	**100°F**
10.	32°C	**90°F**

Exercise C

	Celsius	Kelvin
11.	0°C	**273 K**
12.	100°C	**373 K**
13.	50°C	**323 K**
14.	–20°C	**253 K**
15.	75°C	**348 K**

16. **A** is **correct** because 1 calorie (small c) is the amount of heat energy needed to raise 1 g of water by 1°C.

    B is incorrect because 1,000 calories (also known as a kilocalorie) will raise 1,000 g of water by 1°C.

    C is incorrect because watts are units of electrical power.

    D is incorrect because watts are units of electrical power.

17. **A** is incorrect because temperature is a measure of the kinetic energy in an object.

    B is incorrect because temperature in Kelvin is still a measure of kinetic energy.

    C is **correct**.

    D is incorrect because "average kinetic energy" just means temperature.

18. **A** is **correct** because the heat energy is being used to break the chemical bonds to allow a solid to be a liquid.

    B, **C**, and **D** are incorrect because the temperature remains the same during a phase change.

19. **A** is incorrect because subduction is one tectonic plate sliding under another.

    B and **C** are incorrect because conduction does not work in the emptiness of space.

    D is **correct** because radiation is the only heat transfer method that works through the vacuum of space.

20. **A** is incorrect because subduction is one tectonic plate sliding under another.
    B is incorrect because conduction only works between objects that are touching.
    C is **correct.**
    D is incorrect because radiation is not as effective as convection.

Answers Explained Lesson 7–3

Exercise A

	Millibars	"Station Model"
1.	1028.0	**280**
2.	1008.4	**084**
3.	992.2	**922**
4.	976.6	**766**
5.	994.8	**948**
6.	1000.1	**001**
7.	1008.2	**082**
8.	987.1	**871**
9.	988.8	**888**
10.	1022.2	**222**

Exercise B

	Millibars	"Station Model"
11.	**1028.1**	281
12.	**1020.6**	206
13.	**1008.0**	080
14	**1016.8**	168
15.	**980.0**	800
16.	**988.8**	888
17.	**1016.5**	165
18.	**976.8**	768
19.	**1000.0**	000
20.	**1098.7**	987

21. The higher the pressure is, the clearer the weather. The lower the pressure the stormier the weather.

22. **A** is incorrect because isotherms connect equal temperatures.
    B is **correct**. *Bar* in "isobar" comes from *bar*ometric pressure.
    C is incorrect because isosceles is a type of triangle.
    D is incorrect because contour lines connect equal elevations.
23. **A** is **correct** because, if the pressure is increasing towards the center then the center has the highest pressure around.
    B is incorrect because the values are increasing towards the center.
    C is incorrect because the center is a high pressure area. High pressure has calm and clear weather.
    D is incorrect because the center is a high pressure area. High pressure centers are clear.
24. **A** is incorrect because pressure is the weight of the air above you. On the top of a mountain, there is less air and there is less pressure.
    B is incorrect because air pressure decreases as you go higher.
    C is **correct** because the tops of mountains have less air above them pushing down, creating less pressure.
    D is incorrect because it is a well-established relationship between altitude and air pressure.

Answers Explained Lesson 7–4

Exercise A

	Dry Bulb	Wet Bulb	Relative Humidity	Dew Point Temp.
1.	12°C	7°C	**48%**	**1°C**
2.	22°C	20°C	**83%**	**19°C**
3.	18°C	12°C	**48%**	**7°C**
4.	6°C	5°C	**86%**	**4°C**
5.	22°C	**21°C**	**92%**	21°C
6.	20°C	**16°C**	66%	**14°C**
7.	19°C	19°C	**100%**	**19°C**
8.	17°C	13°C	**63%**	**10°C**
9.	26°C	24°C	**85%**	**23°C**
10.	16°C	12°C	**62%**	**9°C**

Exercise B

11. **evaporation**
12. **evaporation**
13. **condensation**
14. **condensation**
15. **kinetic energy**
16. **condensation**
17. **evaporation**
18. **melting**
19. **freezing**
20. **sublimation**

Answers Explained Lesson 7–5

1. **B** is the **correct** answer because with a steeper pressure gradient there will be stronger wind. They both go up so it is a **direct** relationship.
2. **C** is **correct** because low pressure centers suck the air inward and it spins counterclockwise. Use the right-hand rule with the thumb pointing up.
3. **B** is **correct**. It is important to know where the wind is coming from because that is where the weather is coming from.
4. **A** is incorrect because weather maps do not use this symbol.

   B is incorrect because the number in the lower left of a station model is the dew point temperature.

   C is **correct**.

   D is incorrect because the shaft of the arrow shows the wind direction.
5. **C** is correct because we feel lower pressure as the air rises up off of us. As the air rises, it cools and condenses, creating clouds.

Answers Explained Lesson 7–6

1. **A** is **correct** because a warm front has gradual rising of the air, which makes an overcast sky with a little precipitation falling.

   B is incorrect because this is the description of a cold front.

   C is incorrect because the passage of a warm front means that the temperature is changing.
2. **A** is **correct** because maritime refers to "ocean" and tropical refers to equatorial regions.

   B is incorrect because over land and closer to the equator is describing a continental tropical air mass.

C is incorrect because a maritime polar air mass would form over cold water.

D is incorrect because a continental polar air mass would form over cold land.

3. **B** is **correct** because, in front of the front, the cold air hasn't moved in yet.
4. **A** is **correct** because a warm front is wedge-shaped and the clouds form in front of the front.

   B is incorrect because this is the way that a cold front behaves.

   C is incorrect because, after the front passes your location, weather conditions become clear and stable.

   D is incorrect because the location of rain compared to the position of the front depends on the direction that the front is moving. A certain compass direction is not associated with it.
5. **A** is **correct** because cold fronts produce rapidly rising air, which creates towering thunderstorms.

   B is incorrect because a warm front will have gentle rain.

   C is incorrect because a stationary front is not consistent in its behavior.

Answers Explained Lesson 7–7

1. **A** is **correct** because the winds move across the United Stated from the west to the east.

   B is incorrect because the horse latitudes is a band in between two wind belts where it is very calm.

   C is incorrect because the United States is not in the tropics. It is in the mid-latitudes.

   D is incorrect. Even though the United States is surrounded by water on the east and west, it has little influence on the overall prevailing winds.
2. **B** is **correct**. As the air pressure drops, the pressure gradient between the center of the storm and the outside gets steeper. The wind gets sucked in faster.
3. **A** is incorrect because a land mass is not a source of moisture that the hurricane needs. Land masses actually weaken hurricanes.

   B is incorrect. Even though a hurricane is an atmospheric disturbance, it doesn't get its power from the air.

   C is **correct** because the warmth adds energy and the evaporated water supplies the moisture. In addition, as the water condenses into clouds, more energy is released, adding even more strength to the storm.

   D is incorrect because lightning does not supply energy to a hurricane.

4. **A** is incorrect because a thunderstorm is a localized storm. It is not a low pressure system.

   B is incorrect because a hurricane is a tropical cyclone. It is a low pressure system but it did not originate in the mid-latitudes.

   C is incorrect because a tornado is a localized storm. It is not a low pressure system.

   D is **correct** because a nor'easter is a low pressure system that originated in the middle latitudes.

5. **A** is incorrect because winds by themselves are not very damaging. It gets dangerous when the winds start to pick up materials.

   B is incorrect. Even though wind-blown objects are dangerous and do kill people in hurricanes, it is not the biggest killer.

   C is **correct** because most people who die in hurricanes die by drowning. They often do not evacuate coastal areas early enough and get washed away as the storm intensifies.

   D is incorrect. Thunderstorms do get mixed in with hurricanes but, just as with a regular thunderstorm, the odds of a person being struck are remote.

Chapter 7 Exam Answers Explained

Exercise A

Complete the chart below using the relative humidity and dew point temperature charts.

	Dry Bulb	Wet Bulb	Relative Humidity	Dew Point Temp.
1	30°C	21°C	**44%**	**16°C**
2	0°C	–2°C	**63%**	**–6°C**
3	25°C	21°C	**69.5%**	**19°C**
4	11°C	6°C	**45.5%**	**–.5°C**
5	26°C	**19°C**	**51%**	15°C
6	8°C	**1°C**	17%	**–14°C**
7	20°C	**16°C**	**66%**	14°C
8	16°C	**8°C**	**29%**	–1°C

Exercise B

	Temp °F	Dew Pt. Temp °F	Air Press. mb.	Sky Cover	Wind Speed	Wind Direct.
9. 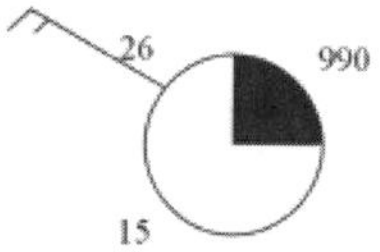 26 990 15	**26**	**15**	**999.0**	**25%**	**15kts**	**nw**
10. 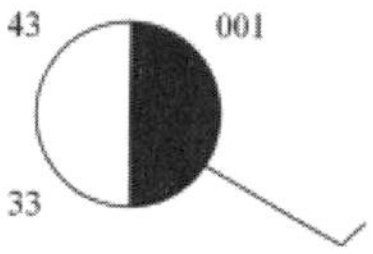 43 001 33	**43**	**33**	**1000.1**	**50%**	**10 kts**	**se**
11. 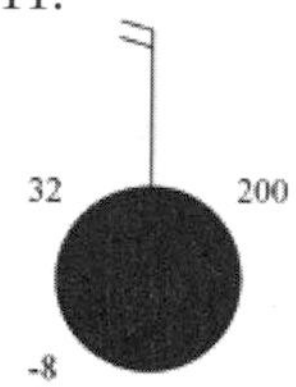 32 200 -8	**32**	**–8**	**1020.0**	**100%**	**20 kts**	**n**
12. 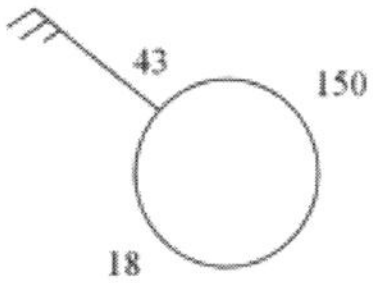 43 150 18	**43**	**18**	**1015.0**	**0%**	**25 kts**	**nw**
13. 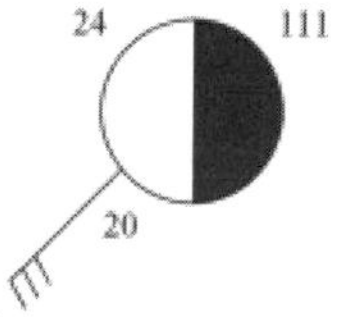 24 111 20	**24**	**20**	**1011.1**	**50%**	**30 kts**	**sw**

Exercise C

14. **6 a.m.**
15. **6 a.m.**
16. **3 p.m–4 p.m.**
17. **2 p.m.–3 p.m.**
18. As the air temperature increases, the relative humidity increases. This is because hotter air can hold more moisture.
19. **6 a.m.**
20. **100%**
21. **6 a.m.** The relative humidity was 100% and the dew point and air temperatures were equal.
22. **3 p.m.** Air temperature highest and relative humidity was lowest.

Exercise D

23.

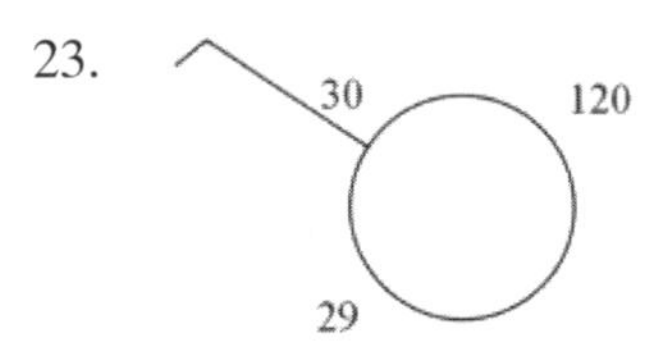

24.

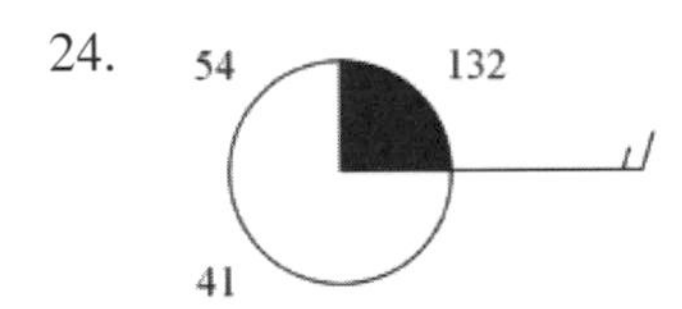

25.

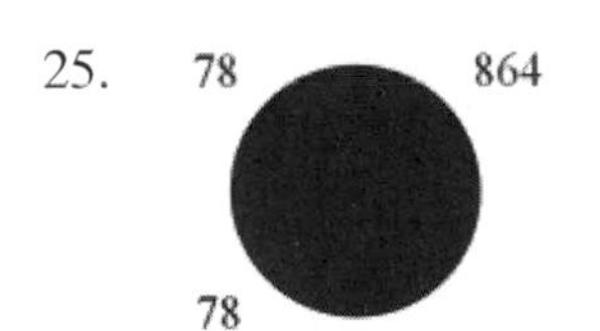

26.

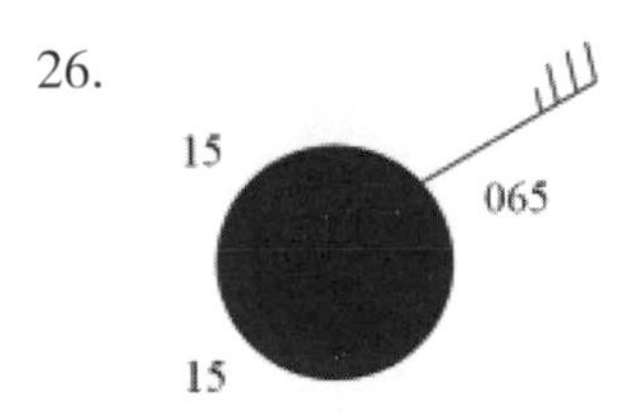

27.

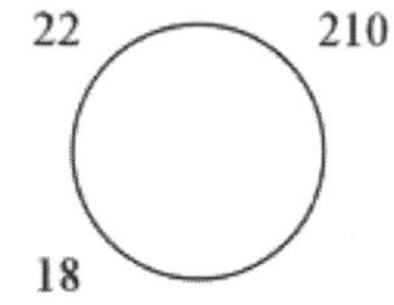

Exercise E

28. **True**.
29. **False**. **Warm** fronts are usually associated with long periods of gentle rain.
30. **False**. When the air temperature is very close to the dew point temperature the air is very **humid/moist/wet**.
31. **True.**
32. **False.** A description of weather conditions at a particular weather station is called a **station model**.

The Water Cycle and Climate

Lesson 8–1: The Water Cycle

The **water cycle** (also known as the *hydrologic cycle*) describes the movement of water around our planet. The majority of the water on the Earth is the same water that has been here since Earth's infancy, and it will be the same water for millions of years into the future. There are very few natural processes that create or destroy water.

The water originally came from a combination of two sources: steam escaping from early active volcanoes and ice melted from comets colliding into the young Earth. However, once the water got here, it stayed. Since then, the water has been recycled countless times through the water cycle.

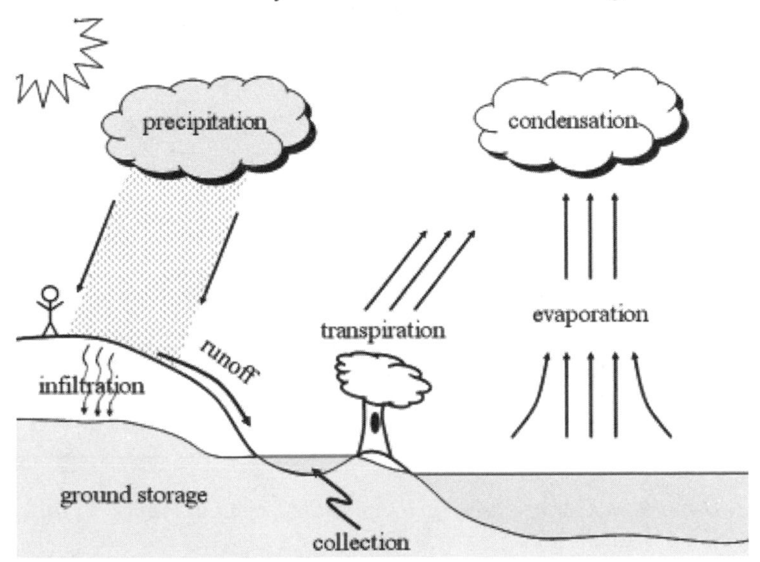

Figure 8.1. Water cycle

The diagram of the water cycle does an excellent job explaining how the system works, but there are a few terms that need clarification:

Transpiration is a biologic activity of plants. Water escapes from the leaves of plants and enters the atmosphere. In heavily vegetated areas this can be a significant source of atmospheric water. Sometimes, the terms *evaporation* and *transpiration* are lumped together into one word **evapotranspiration**. This simply encompasses all of the possibilities of water vapor coming from land sources.

Groundwater

When water hits the ground there are a few things that the water can do. It can evaporate, collect, run off, or infiltrate.

When water collects on the surface of the land it produces standing water, puddles, ponds, or lakes. Runoff is the term for water that slides over the surface looking for the lowest points around—in other words: rivers.

Infiltration is the process that lets water get into the ground. The amount and rate of infiltration depends on how much space there is inside the ground and how well the spaces are connected. Water that seeps into the ground fills in spaces between the sediments called **pores**. **Porosity** is a measure of the percentage of the rock material that is made of empty spaces and is calculated with the equation:

$$\text{porosity} = \frac{\text{volume of pore space}}{\text{total volume of soil}} \times 100$$

Porosity can be affected by the sediment's shape, sorting, and packing.

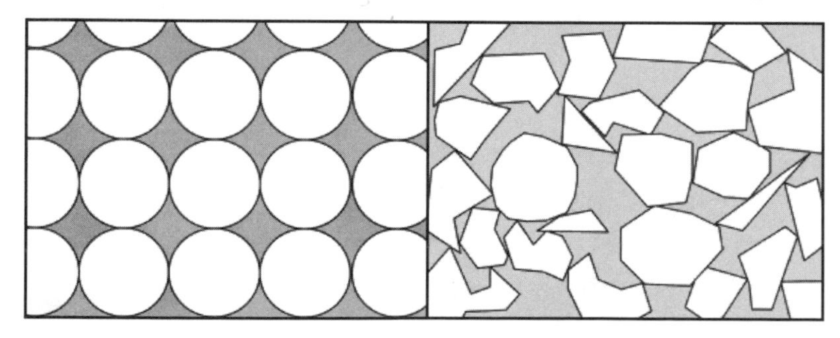

Figure 8.2. The spherical shapes on the left and the irregular shapes on the right affect the soil's porosity.

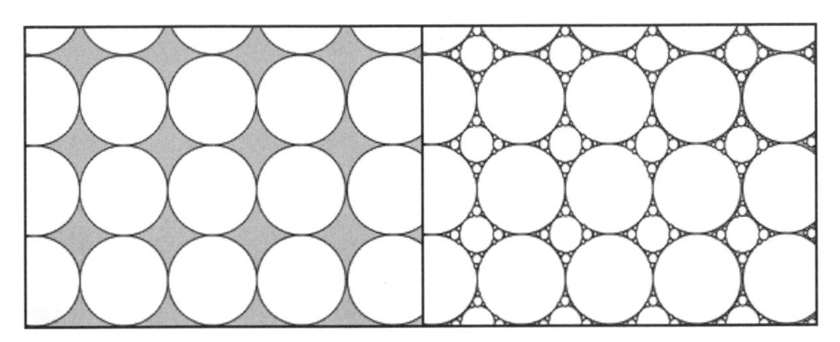

Figure 8.3. The sorted soil on the left has a higher porosity than the mixed sediments on the right.

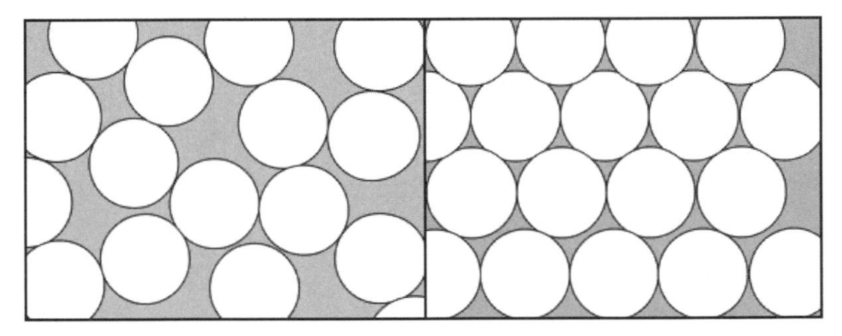

Figure 8.4. The packing affects the amount of pore space.

One thing that does not affect porosity is the size of the particles. Imagine sediments made of spheres. The ratio of solid material compared to the spaces between them will stay the same regardless of the size of the spheres. Smaller particles will have smaller pores, but a lot more pores. Large sediments have few spaces, but the spaces take up a large volume.

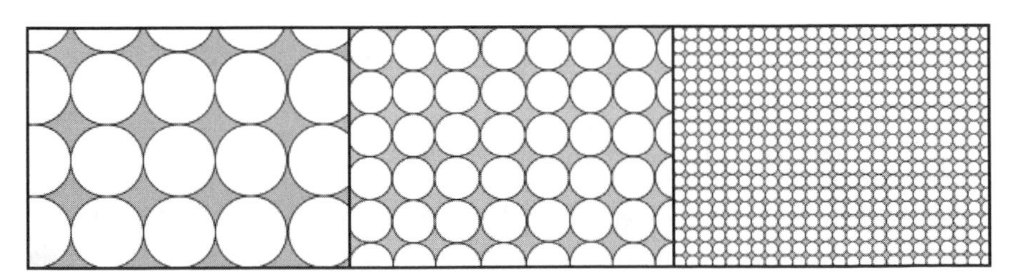

Figure 8.5. The ratio of sediment to empty pore space is the same in all three diagrams.

Another factor to consider, when explaining how much water will infiltrate into the ground, is the ability of water to penetrate into the soil. Even though porosity is a major factor, the pores must be connected to allow flow. The ability of water to flow through rock or soil is called **permeability** (the ability to permeate). Just because a material has a high porosity, it does not automatically mean that it has a high permeability. Volcanic pumice is a prime example of a rock with high porosity but low permeability. Pumice is a bubbly rock made mostly of gas pockets trapped within the rock. Pumice floats! It floats because the gas pockets are completely sealed and act as little life preservers. The same feature that allows pumice to float also forbids water from permeating the rock: The chambers are sealed and the water cannot penetrate. Conversely, some very solid, non-porous igneous rocks can have a high permeability. This is accomplished if the rock is fractured, which may provide channels for the water to flow through. (This is sometimes cited as a cause of water levels changing shortly before earthquakes. The stress from the impending quake fractures bedrock and allows groundwater to redistribute.)

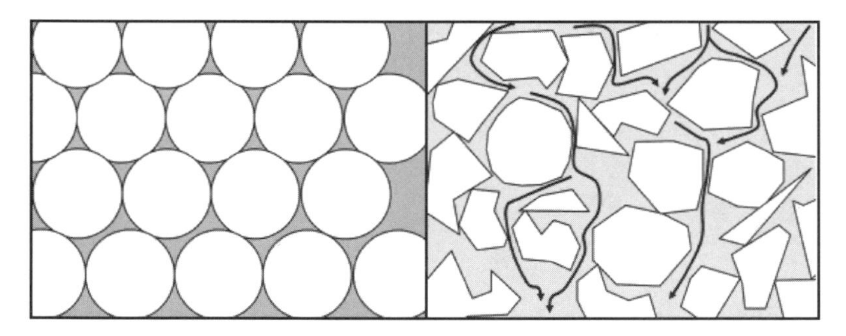

Figure 8.6. The sample on the left has a very low permeability because the pores are not lined up. In the sample on the right, the water easily and quickly flows through the sediments.

Lesson 8–1 Review

1. Evapotranspiration is a combination of which two processes?
   a) evaporation and transportation
   b) evaporation and elevation
   c) evaporation and transpiration
   d) evaporation and condensation

2. Infiltration is the process of water
   a) evaporating into the air.
   b) seeping down through the soil.
   c) getting sucked upwards against the pull of gravity through tiny pores.
   d) filling in all of the pore space in soil.
3. The porosity between large spheres will be
   a) greater than that of small spheres.
   b) less than that of small spheres.
   c) equal to that of small spheres.
   d) not enough information.
4. Which statement best describes why the volcanic rock pumice floats?
   a) Pumice has a high porosity but a low permeability.
   b) Pumice has a high porosity and a high permeability.
   c) Pumice has a low porosity and a low permeability.
   d) Pumice has a low porosity but a high permeability.

Lesson 8–2: Groundwater

Water will infiltrate into the ground as long as there are pores and the ground is permeable and not already filled with water. The materials (sediment or rock) that water can reside in—or flow through—are called **aquifers**.

In the simplest situation, water within the ground will rest upon the top of solid bedrock and "pile up" to make a zone of saturated soil. **Saturated soil** has all of its pores completely filled with water. The top surface of this subterranean water storage is called the **water table**. The water table changes its height depending on rainfall and weather, as well as the amount of human pumping for drinking and agriculture. The water table does not have a flat surface on its top. The table mimics the shape of surface topography—that is it gets higher in hills and lower in valleys. In areas where the level of the land falls below the water table, there will be *surface water* present. This can be in the form of a lake, swamp, or river.

Sitting on top of the water table is a zone where some of the water gets drawn upward against the pull of gravity by a "sucking" effect of the pores themselves called **capillarity**. And this area is called the **capillary fringe**. Capillarity is an effect that gets its name from the tiniest blood vessels in our bodies: the capillaries. When water gets into tiny spaces, it clings the sides and climbs upward a little—a process called **adhesion**. This is also the cause of the curved surface at the edges of a glass of water. The effect works best in smaller spaces, so the smaller the pores, the greater the capillarity. Therefore, soils and rocks made from smaller sediments have smaller pores and, as a result, a thicker zone of capillarity.

Above the zone of capillarity is an area where the pores are not saturated with water. Some water will still cling to the side of the sediments inside the pore but the pores are essentially filled with air. This is called the **zone of aeration**.

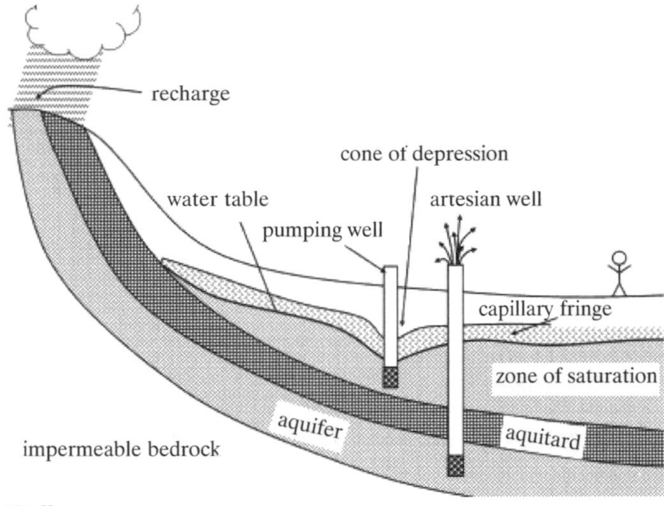

Figure 8.7. Wells

To dig a simple well, a hole has to be dug at least down to the water table. However, as conditions change, the water level can drop and the well will dry up. For a reliable well, the hole must be dug below the lowest level that the water drops to throughout the seasons.

A **spring** is a spot where the water table emerges from the ground and the water flows out.

An **artesian well** is the more dramatic well where the water shoots upward under natural pressure. In order to have an artesian well a few circumstances must come together. The **aquifer** containing the water must be sandwiched between two impermeable (or at least less permeable) layers called **aquitards**. Next, the three layers must be tilted at an angle. Finally, the source of water at one end must be higher than the point where the well is located.

The water table can be drastically influenced by human activity. In the immediate area in which water is being pumped from a well, the water table will sink lower around the bottom of the pipe, making a feature called a **cone of depression**.

In populated areas where groundwater is used extensively it becomes difficult to put the water back into the ground for future use. If too much of the open ground is covered with pavement such as roads and sidewalks, then much of the water that should have gone into the ground gets diverted to other areas. The way that this is compensated for in many communities is to have recharge basins scattered throughout the area. A **recharge basin** is a large pit lined with sand and gravel. The pit collects rain water and storm runoff and allows it to infiltrate back into the groundwater.

Another problem that populated communities in coastal areas must deal is salt water intrusion. Saltwater from the ocean seeps into the aquifer when the fresh water in the aquifers is over-pumped. Wells near the coast will start to draw saltwater instead of fresh.

It is possible to have a fresh water supply on a sandy island surrounded by sea water. Fresh water is less dense than saltwater. If the two are placed together very carefully, the fresh water will float on top of the denser salt water. This creates a fresh water lens that could supply drinking water.

Lesson 8–2 Review

1. Layers underground where water can naturally flow are called
   a) the capillary fringe.
   b) aquifers.
   c) pore space.
   d) an artesian well.

2. When the ground is filled to its capacity with water it is
   a) humid.
   b) packed.
   c) aerated.
   d) saturated.

3. Which of these processes does not cause water to go upward?
   a) capillary action
   b) infiltration
   c) evaporation
   d) transpiration

4. An artesian well
   a) flows under its own pressure.
   b) needs to be pumped.
   c) must be higher than its source.

5. Saltwater intrusion is caused by
    a) too many recharge basins in an area.
    b) excessive rainfall.
    c) fresh water lenses.
    d) over-pumping near a coast.

Lesson 8–3: Water Budget

The **water budget** is a mathematical model that is used to describe the type of climate that an area has. It essentially puts numbers on how much water is used and how much is available. The analogy of a financial budget is inescapable and actually makes it easier to understand. Here is a chart with the terms of the water budget and their counterparts in a money budget. More detailed descriptions of the difficult terms follow.

Symbol	Water Budget Term	Money Budget Term
P	Precipitation: how much water falls in a given month	Pay: how much money you get in a month
E_p	Evapotranspiration Potential: strength of the sunlight	Bills: must be paid if you have money
$P - E_p$	Net income of water	Take home pay: the amount of money you have left after paying bills
E_a	Actual Evapotranspiration: amount of water that actually evaporates	Paid bills: amount that you actually paid out
St	Storage: amount of water saved in the ground	Savings account: balance of your savings
ΔSt	Change in storage: amount of water that is put into storage or taken out	Deposits and withdrawals
$+\Delta St$	Recharge: water is being put back into ground storage	Deposit: money is put into savings
$-\Delta St$	Usage: water is taken out of the ground.	Withdrawal: money is taken out of the bank to pay bills
D	Deficit: there's not enough water	Debt: you used up all of your pay and savings and still have bills
S	Surplus: extra water that turns into runoff	Shopping spree: you run off to the mall

E_p and E_a

The two most misunderstood terms are E_p (Evapotranspiration Potential) and E_a (Actual Evapotranspiration). A simple way of thinking of E_p is to just think of it as "the strength of the sunlight" because that is what determines E_p. E_a is always going to be the same as E_p unless you've run completely out of water in precipitation (P) and storage (St.).

The two systems work almost exactly the same with the exception of a couple of extra rules. If you have bills, you must pay them even if you have to empty your bank account. Your bank can only hold 100 dollars (100 mm of water). If you have any more than that, a surplus, you must get rid of it (run off to the mall).

Here is a step-by-step guide to filling out a water budget:

- Fill out your P and E_p.
- Subtract E_p from P ($P - E_p$) to see how much money you have to work with. This will be your take-home pay after paying bills.
  - If you still have money left over, you put it in the bank.
  - If you don't have enough pay to pay your bills, take money out of the bank.
- ΔSt is any transaction you do at the bank. (Δ means "difference.") A negative ΔSt is a withdrawal, and a positive is a deposit. You can only go up to a maximum of 100 or a minimum of 0 in the bank. Any more than that goes into surplus and any less is a deficit.
- E_a is how much of the bills that were paid including any that you used from storage.

The water budget is used by civic planners and farmers to prepare for times of deficit as well as surplus. Obviously, a time of deficit is dry and farmers need to prepare to irrigate their crops. However, during times of surplus, civic planners must be prepared to drain off the extra water or risk flood.

Lesson 8–3 Review

1. A deficit exists when
   a) $P - E_p$ is negative and storage is zero.
   b) E_a is higher than E_p.
   c) there is too much rain in an area.
   d) you have a period of usage.
2. Surplus water will
   a) be used to recharge ground storage.
   b) be used to "pay the bills."
   c) cause a deficit.
   d) flow over the surface of the land.
3. Maximum groundwater storage for most areas is about
   a) 10 mm. b) 50 mm. c) 100 mm. d) 100 gallons.
4. Usage will happen when
   a) E_p is higher than P.
   b) E_p is lower than P.
   c) E_a is higher than E_p.
   d) St is greater than 100.

Lesson 8–4: Location and Climate

Latitude

Probably the most basic relationship between location on the planet and your climate has to do with the latitude of the region. The Earth is heated by the Sun's radiation. If you are in a place where the radiation is strongest, it will be hotter. The equator is reputed as being the place on Earth where the Sun's rays are strongest. That is not exactly true, but it is a good approximation. The Sun's rays strike the equator at close to a 90° angle all year long. The angle at which the Sun's light strikes the ground is called the **angle of insolation** (not to be confused with ins*u*lation, which is a material that is used to keep in heat). Insolation is a combination of the words "*in*coming *sol*ar radi*ation*." The closer the angle of insolation is to 90°, the stronger the Sun's light and the warmer it will be. Think of it this way: If you take a flashlight and point it straight down onto a table, there will be a small compact area where all of the energy is being concentrated. Now if you shine the light at a glancing angle, the circle of light, as well as all of the energy, gets spread out all the way across the table. The Sun's light is strongest near the equator and weakest at the poles. This explains why it is warmer near the equator and, the further away you get, the colder the climate gets.

Elevation

As discussed in Chapter 5, temperature gets colder the higher up you are. As a result, if you have a city that is built at a high elevation, it will have a cooler climate than the lower areas at the same latitude. This is why there can be snow-capped mountains near the equator. The peak of Mona Loa in Hawaii has snow on its peak for most of the year!

Coastal and Continental Climates

Water is stubborn and doesn't want to change its temperature. If you live near a large body of water, this can have a major impact on your climate. Because water doesn't want to change its temperature, an area with a lot of moisture will have very stable temperatures. This is called moderating the temperature. On the other hand, a dry climate will have wildly swinging temperatures. Deserts are reputed as having extremely hot days and nights that are just as extremely cold.

The interior of the United States changes temperatures very quickly because it is so far from major bodies of water. Coastal areas still have changes in temperature, but they are less extreme and take longer to happen. The delay in temperature is called the **time-temperature lag** and is a direct result of water's high specific heat ("stubbornness to change"). You would expect the hottest time of the day to be at 12 noon, but it is not. It takes a little while for the moisture in the atmosphere to warm up. As a result the hottest time of the day is around 2 p.m. Likewise, the hottest time of the year is not on June 21, but closer to the end of July. This is due to the time that it takes the entire ocean to heat up.

Ocean Currents

Ocean currents are large circulations in the waters of the world's oceans. Some of these currents are truly global. The currents' purpose is to try to equalize the planet's water temperature. Cold currents flow from the poles to the equator and from the equator to the poles. Currents just off the coast will bring along with them the heat or cold of their source region. For example, England is at the same latitude as the middle of Canada, but it has a much warmer climate. That is because the Gulf

Stream current carries warm Caribbean waters to the shores of England. This also explains why England has the reputation of being so foggy. Warm, moist air collides with the colder latitudes, producing fog, much the same way that fog comes out of the refrigerator when the cool air from inside collides with the warmth of the room.

Lesson 8–4 Review

1. A coastal climate will have
   a) very warm temperatures.
   b) very cold temperatures.
   c) highly variable temperatures.
   d) very stable temperatures.
2. As the latitude of a region increases (gets closer to 90°),
   a) the altitude will also increase.
   b) the climate will get colder.
   c) the climate will get warmer.
   d) the strength of the Sun will get stronger.
3. The angle of insolation is the
   a) time difference between the strongest sunlight and the hottest time.
   b) latitude of the observer.
   c) angle at which the Sun's light strikes the ground.
   d) angle to Polaris.
4. The hottest time of the year is
   a) around June 21st.
   b) at the end of July.
   c) about 2 in the afternoon.
   d) happens whenever the Sun is at its highest point in the sky.
5. It is possible to snow at the equator if
   a) you are surrounded by water.
   b) you are at a high enough elevation.
   c) the angle of insolation is high.
   d) you remain at sea level.

Lesson 8–5: Mountain Barriers

The most obvious way to change the temperature of air is to heat it up or cool it by adding or removing heat. There is another way that a region of air, called a parcel, can change its temperature. Simply put, the heat can get concentrated, making the air feel hotter, or the heat can get spread out, making the air feel cooler. When a parcel of air rises, it is surrounded by less air and expands as a result of the lower pressure. When the air expands, the atoms of gas, along with their heat, spread out. As the heat is spread out, the temperature lowers. This process of cooling air without removing heat is called **adiabatic cooling**.

Adiabatic cooling can be observed whenever a pressurized canister is quickly emptied. Frost will often form on the outside of the can. Adiabatic cooling also explains why the tops of mountains are generally colder than the valleys and is the reason why clouds will form high above drier air. As the air rises, it cools, which lowers the dew point temperature. Also, while the cooling air's capacity for holding moisture drops, the relative humidity increases. If the air continues to rise, the dew point will get closer to the air temperature and the relative humidity will get closer to 100%. At this point clouds will form (assuming condensation nuclei are present).

The opposite process applies: As air sinks, the atoms get closer concentrating the heat. The concentrated heat makes the air hotter, raising the capacity for holding moisture. The air gets drier.

If Hot Air Rises, Why Is it Colder High Up?

It is true that hot air rises. But as the air rises, it expands and cools by adiabatic cooling. This makes the air colder. If the air gets colder, then doesn't it start to sink again? In this case, it does not. The air will continue to rise as long as it is surrounded by air that is warmer—even just a little warmer. Eventually, mixing with surrounding air or some kind of barrier (temperature inversion) will stop the rising air.

If cold air sinks and heats up as it sinks, why are high pressure areas usually colder? The air that descends heats up through adiabatic heating, yet it will continue to sink because it is still cooler than the surrounding air.

The Orographic Effect (or the Mountain Barrier Effect)

When a weather system encounters a mountain range, it gets forced to rise and go over the range. This will have a profound affect on the climates on either side of the mountains. This is called the Orographic Effect (*ora-* refers to the "mountains" as in orogeny). The side of the mountain that gets hit by the incoming weather is called the **windward** side. On the windward side of the mountain, the air is forced to rise up. As it rises, the pressure lowers and the air expands and cools because of adiabatic cooling (heat gets spread out). The lowering of the air's temperature also lowers the air's ability to hold any water that it is carrying—the relative humidity increases. If the rising of the air continues, the relative humidity will continue to increase until condensation occurs, creating clouds and eventually precipitation. As a result, the windward sides of large mountain ranges are often very wet. A prime example is the temperate rain forests located in the states of Washington and Oregon.

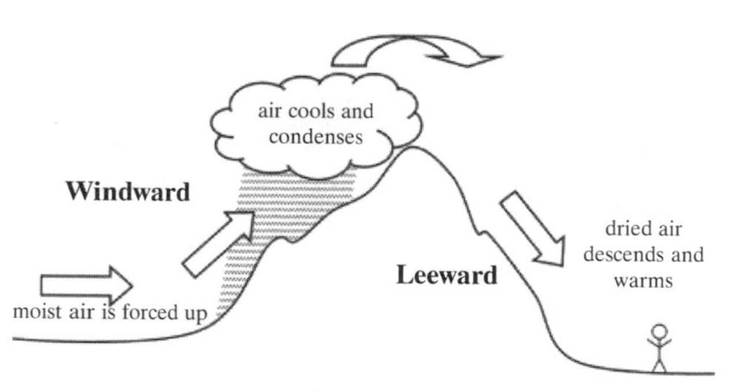

Figure 8.8. Orographic Effect

The precipitation on the windward side of the mountain drops much of the water out of the air before the air crosses the mountains and goes down the other side. On the other side of the mountain, called the **leeward** side, the newly dried air begins to descend. As it descends, it experiences higher pressure, the air compresses, and adiabatic heating raises the temperature. The increased temperature raises the capacity of the air to hold more moisture so the relative humidity decreases. The air on the leeward sides of mountain ranges is often very dry. The leeward sides of mountain chains are called **rain shadows**. The condensation that happened on the windward side released a little heat into the air (evaporation removes heat, and condensation adds heat), and the descending air on the leeward side will now be a few degrees warmer. The warm, dry breeze that comes down off of a mountain range is called a **Chinook**.

Lesson 8–5 Review

1. The bottom of the leeward sides of large mountains tend to be
   a) moist. b) cold. c) dry. d) cloudy.
2. When air sinks it will
   a) expand and cool.
   b) expand and heat.
   c) contract and cool.
   d) contract and heat.
3. The side of the mountain that gets hit by the wind is called
   a) the windward side.
   b) the leeward side.
   c) the orographic side.
   d) a Chinook.
4. Which of the following describes adiabatic cooling?
   a) Air cools because the angle of insolation decreases.
   b) Air cools because heat is removed.
   c) Air cools because the volume increased and spread the heat out.
   d) Air cools because it is closer to space.

Lesson 8–6: Land Breeze/Sea Breeze

The shore is an interesting climate because it is the contact of two very different environments: the water and the land. Water has a very high specific heat, which means that water heats up very slowly and also cools off very slowly. When it comes the changing its temperature *water is very stubborn*. By contrast, the land will heat up and cool down rather quickly.

If we explore this system over the course of a day, we can see some stark differences between how it works during the day and during the night. In the span of a single day, the water's temperature changes so little that it can be ignored. The land, however, changes its temperature drastically.

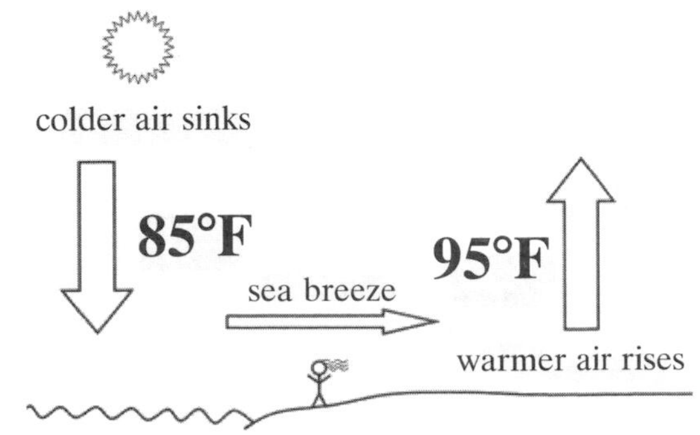

Figure 8.9 Sea breeze

In the heat of the midday Sun, the land heats up quickly. At the shore, the contrast of hot land right next to cooler water sets up convection. The hot air over the land rises, creating low pressure, while the cool air over the sea sinks, creating high pressure. The air in between flows from the sea to the land (high pressure to low pressure), making a **sea breeze**. Standing on the shore and watching the ocean will put the wind on your face at this time. If the convection is strong enough, it could create a line of puffy cumulus clouds over the land and paralleling the shore.

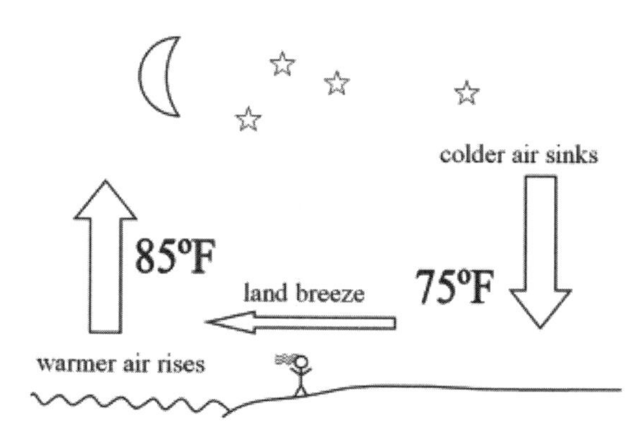

Figure 8.10. Land breeze

As the Sun begins to set in the afternoon, the land cools until there is less and less of a temperature difference between the sea and the land; convection slows and stops.

In the evening the land continues to cool until it eventually is cooler than the stable temperature of the ocean. The convection cell starts up again but in the opposite direction now. The sea water, by contrast, is now warmer than the land. The warmer air over the sea rises and the cooler air over the land sinks. This sets up a breeze going from the land out to sea: a **land breeze**.

Lake Effect

A similar effect to the land breeze/sea breeze situation is *lake effect snow*. When prevailing winds go across a large lake, they pick up moisture. Upon reaching the far shore, the moist air will rise during the day just like at the beach. The rising air will cool and condense. If this happens during the winter, when the temperatures are below freezing, areas that have lake effect snow often get frequent, heavy snowfalls.

Lesson 8–6 Review

1. The specific heat of water is very high. This means that water
   a) heats up very quickly.
   b) only evaporates at high temperatures.
   c) heats up very slowly.
   d) is easy to warm up.

2. In a shore environment the land will
   a) heat up faster than the water.
   b) heat up at the same rate as the water.
   c) heat up slower than the water.
   d) cool down slower than the water.
3. During the day a sea breeze will blow
   a) onto the shore.
   b) out to sea.
   c) towards sinking air.
   d) towards the higher pressure.
4. During a land breeze, the lower pressure will be over the
   a) and.
   b) breaking waves.
   c) sea.
   d) mountains.

Chapter Exam

1. As we all know, hot air rises. But as the air rises, what will happen to the temperature of the hot air?
   a) The warm air will cool down.
   b) The warm air will stay the same temperature.
   c) The warm air will continue to heat up.
   d) The pressure will increase.
2. A desert's water budget will operate in which mode for all or most of the year?
   a) surplus b) deficit c) recharge d) storage
3. How would the angle of insolation vary throughout the day in one location?
   a) It would increase.
   b) It would decrease.
   c) It would stay the same.
   d) It would increase and then decrease.

4. At which point in the water cycle is new water created?
   a) evaporation
   b) condensation
   c) precipitation
   d) none of the above
5. Why does it feel warmer when it is cloudy at night?
   a) The clouds trap in heat.
   b) Night clouds are filled with warm water.
   c) The clouds are made from low pressure, which is warm, rising air.
   d) The clouds, focus starlight through them.
6. If warm air rises, then how does it make clouds which require cold temperatures?
   a) Adiabatic heating causes the water to steam out of the air.
   b) The rising air picks up condensation nuclei.
   c) The air expands, which spreads out the heat and it gets cooler.
   d) It rises into the jet stream, which is cold air.
7. England is warmer than similar latitudes around the planet because
   a) England gets a Chinook from the Alps.
   b) the warm Gulf Stream current flows right into England.
   c) the industrial pollutants in England trap in most of the heat generated in the area.
   d) England is in a low pressure belt, which is warm air.
8. If you traveled from the equator to the North Pole, how would your angle of insolation change?
   a) It would increase.
   b) It would decrease.
   c) It would remain the same.
   d) It depends on your longitude.
9. What is the relationship between particle size and porosity?
   a) Larger particles have greater porosity.
   b) Smaller particles have greater porosity.
   c) Mixed particles have greater porosity.
   d) Size does not affect porosity.

10. Which situation describes a material with low porosity but high permeability?
    a) a vesicular (bubbly) piece of igneous pumice that can float.
    b) a solid concrete foundation of a house.
    c) loose sand and gravel lining the side of a road.
    d) a solid piece of bedrock that has been fractured by earthquake activity.

11. Water being drawn upward against the pull of gravity is called
    a) porosity. b) capillarity. c) permeability. d) saturation.

12. A sea breeze will last
    a) as long as the Sun shines brightly.
    b) for an entire day.
    c) until the next weather system comes in.
    d) until the next low tide.

13. As you travel from the North Pole to the equator, the potential for evapotranspiration (E_p) will
    a) increase.
    b) decrease.
    c) remain the same.
    d) increase and then decrease.

14. Insolation means
    a) incoming solar radiation.
    b) infrared radiation.
    c) solar evaporation.
    d) hydrologic condensation.

15. Where would you expect to find a desert climate with respect to a mountain chain?
    a) on the windward side of the mountain
    b) at the top of the mountain
    c) on the leeward side of the mountain

16. Groundwater is stored within the
    a) zone of aeration.
    b) aquitards.
    c) aquifers.
    d) cone of depression.

Answer Key

Answers Explained Lesson 8–1

1. **C** is **correct**. Evapotranspiration is a combination of evaporation and transpiration.
2. **A** is incorrect because water evaporating into the air is evaporation.

   B is **correct**.

   C is incorrect because water getting sucked upwards against gravity is capillarity.

   D is incorrect because water filling in all of the pore space is saturation.
3. **C** is **correct** because the ratio between the amount of pore volume to the volume of the overall volume of the soil stays the same.
4. **A** is **correct**. Pumice is made mostly of air, which gives it a lot of pore space. The bubbles are totally closed, which doesn't allow water to pass through, which gives pumice a low permeability.

Answers Explained Lesson 8–2

1. **A** is incorrect because the capillary fringe is a section just on top of the water table where water creeps upward.

   B is **correct** because aquifers allow water to flow through the ground.

   C is incorrect because pore space is just an opening. The openings must also be connected to allow permeability.

   D is incorrect because an artesian well is a special situation in which water is pushed to the surface.
2. **A** is incorrect because *humid* is a term used for moist air.

   B is incorrect because packed describes the arrangement of the sediments in the ground.

   C is incorrect because aerated means to have air mixed in the soil.

   D is **correct**. When the ground is filled to capacity, it is saturated.
3. **A** is incorrect because capillary action draws the water upward through tiny pores.

   B is **correct** because infiltration is water seeping down into the ground.

   C is incorrect because evaporation takes water from the ground and puts it into the air.

   D is incorrect because transpiration takes water from plants and puts it into the air.
4. **A** is **correct**. An artesian well is naturally pressurized.

   B is incorrect because artesian wells are under pressure and flow on their own.

C is incorrect because the well must be lower than the source in order for the water to flow.

5. **A** is incorrect because too many recharge basins in an area will replenish the fresh water and drive the saltwater back.

   B is incorrect because excessive rainfall will put more fresh water into the ground, which will push back the saltwater.

   C is incorrect because a fresh water lens is a pocket of fresh water resting on top of salt water.

   D is **correct**. Over-pumping will suck in the saltwater from the edges of the aquifer.

Answers Explained Lesson 8–3

1. **A** is **correct**. A negative $P - E_p$ means that there is not enough precipitation this month to cover how much will evaporate. At this point, storage is used to pay the difference but there is no water in storage.

   B is incorrect because E_a cannot be higher than E_p. The actual evapotranspiration cannot be higher than the potential.

   C is incorrect because too much rain in an area would be a surplus.

   D is incorrect because usage means that storage is being used up. A deficit is after storage is completely gone.

2. **A** is incorrect because a surplus happens after the storage is at its maximum of 100.

   B is incorrect because a surplus happens after the bills are paid and there is still money left over.

   C is incorrect because a deficit is a lack of water, while a surplus is too much water.

   D is **correct**. The ground is saturated and there is nowhere else for the water to go than to sit on the surface and run off.

3. **C** is **correct**. The average groundwater storage for most areas is 100 mm.

4. **A** is **correct.** When the precipitation for the month is not high enough to take care of the potential for evapotranspiration, then water in storage must be used.

   B is incorrect because E_p is lower than P describes a recharge or surplus. "The strength of the Sun is not strong enough to evaporate all of the rain."

   C is incorrect because E_a cannot be higher than E_p.

   D is incorrect because a storage higher than 100 is a surplus.

Answers Explained Lesson 8–4

1. **A** is incorrect because the overall temperatures will be determined by the latitude.

   B is incorrect because the overall temperatures will be determined by the latitude.

   C is incorrect because dry climates have highly variable temperatures.

   D is **correct**. The extra moisture in coastal areas keeps the temperatures stable.

2. **A** is incorrect because altitude depends on the height of the land, not the latitude.

   B is **correct** because, as the latitude gets higher, you get closer to the poles.

   C is incorrect because it gets colder near the poles.

   D is incorrect because the Sun's light is not as concentrated at the poles.

3. **A** is incorrect because the time difference between the strongest sunlight and the hottest time is called the time-temperature lag.

   B is incorrect because the angle of insolation is the complimentary angle to latitude. (Both angles add up to equal 90°.) Essentially they are opposites.

   C is **correct**.

   D is incorrect because the angle of insolation is the complimentary angle to Polaris. (Both angles add up to equal 90°.) Essentially they are opposites.

4. **A** is incorrect because it takes time for the environment to warm up.

   B is **correct** because it takes about a month after the strongest sunlight (June 21st) for the oceans to warm up.

   C is incorrect because 2 p.m. is the hottest time of the day.

   D is incorrect because it takes time for an area to heat up after the Sun has reached its peak.

5. **A** is incorrect because being surrounded by equatorial water will keep the area warm.

   B is **correct** because high elevations are cooler than sea level.

   C is incorrect because a high angle of insolation keeps the area warm.

   D is incorrect because sea level is warmer than high elevations.

Answers Explained Lesson 8–5

1. **A** is incorrect because the moisture has already been dropped on the windward side.

   B is incorrect because the top of the mountain is colder.

   C is **correct**. The moisture has been blocked from coming over the peak of the mountain.

D is incorrect because there is little moisture on the leeward side of large mountains.

2. **D** is **correct**. As air sinks, it encounters more pressure, which squeezes the air molecules together. The denser air has a higher concentration of heat and the temperature goes up.

3. **A** is **correct**.

   B is incorrect because the leeward side is the side away from the wind.

   C is incorrect.

   D is incorrect because a Chinook is a warm, dry wind that comes down the leeward side of the mountain.

4. **A** is incorrect. It is a true statement, but it doesn't answer the question.

   B is incorrect because, the difference between "normal" cooling and adiabatic cooling is that adiabatic cools without changing the amount of heat available.

   C is **correct**. By spreading out the heat, the temperature gets colder while the amount of heat remains constant.

   D is incorrect because there are layers of the atmosphere that actually get warmer as you get closer to space.

Answers Explained Lesson 8–6

1. **A** is incorrect because water heats up very slowly compared to other materials.

   B is incorrect because the specific heat of water has very little to do with the evaporation of water. In addition, water can evaporate at cold temperatures as well as hot.

   C is **correct** because specific heat relates how much energy is needed to raise the temperature. The higher the specific heat, the harder it is to heat up a material.

   D is incorrect because water is difficult to warm up.

2. **A** is **correct** because water heats up slower than any other material.

   B is incorrect because land heats up faster than water.

   C is incorrect because land heats up faster than water.

   D is incorrect because land cools down slower than water.

3. **A** is **correct** because a sea breeze comes from the sea and moves onto shore.

   B is incorrect because a breeze heading out to sea will be coming from the land and will be called a land breeze.

   C is incorrect because wind blows into (or gets sucked into) rising air. When there is a sea breeze the sinking air will be over the sea.

   D is incorrect because wind comes from higher pressure and towards lower pressure.

4. **A** is incorrect because the land will be colder at night. The air will sink over the land and create high pressure.

   B is incorrect because the edge of the shore will have an in-between pressure. The lower pressure will be out to sea while the higher pressure will be over land.

   C is **correct** because the low pressure over the sea will suck the air off the land creating a land breeze.

   D is incorrect because the mountains, which are on land, will have colder temperatures and higher pressure.

5. **A** is incorrect because when wind reaches the windward side of the lake it hasn't picked up the moisture from the lake yet.

   B is incorrect because over the lake, the temperature conditions will be stable and the moist air will not have a reason to rise.

   C is **correct** because on the leeward side, the wind will have already crossed the lake to collect moisture. When it reaches the leeward side, it will encounter warmer temperatures, which will cause it to rise and make snow clouds.

   D is incorrect. Because the wind can come from any direction, lake effect snow depends on the direction that the wind is flowing as compared to where the lake is located. The snow will fall after the wind crosses the lake, regardless of what compass direction that may be.

Answers Explained Chapter Exam

1. **A** is **correct**. As the air rises, it will expand and cool.

2. **A** is incorrect because deserts are dry climates.

   B is **correct** because there is not enough rain in a desert to overcome the potential for evapotranspiration.

   C is incorrect because there is rarely a time in a desert when there is enough water to out match the potential for evapotranspiration.

   D is incorrect because there is rarely a time in a desert when there is enough water to out match the potential for evapotranspiration.

3. **D** is **correct**. The angle starts out very shallow at sunrise. It increases until noon and then it sinks again.

4. **D** is **correct** because the water cycle does not create water. The water in the water cycle is re-circulated over and over.

5. **A** is **correct** because clouds act as insulators, preventing heat from escaping back into space.

   B is incorrect because clouds condense into droplets of cold water or, more often, ice crystals.

C is incorrect. Even though low pressure makes clouds, the warm air cools as it rises.

D is incorrect because starlight is not strong enough to provide any heat to the Earth.

6. **A** is incorrect because adiabatic heating happens when air sinks.

   B is incorrect because just having condensation nuclei is not enough. The air must cool to the dew point temperature.

   C is **correct**. The air will not form a cloud while it is still warm.

   D is incorrect because clouds do not normally rise into the jet stream.

7. **A** is incorrect because England is not near the Alps.

   B is **correct**. The warm Gulf Stream brings some of the warmth of the Caribbean to England.

   C is incorrect because any pollutants that England generates will travel downwind away from the industrialized areas.

   D is incorrect because Canada is in the same wind and pressure belt as England, and Canada has a colder climate.

8. **B** is **correct**. The Sun is most direct near the equator, having an angle of insolation of nearly 90°. As you travel to the North Pole, the Sun would get lower in the sky.

9. **A** is incorrect because larger particles have larger pores but fewer of them. It evens out.

   B is incorrect because smaller particles have smaller pores but much more of them. It evens out.

   C is incorrect because mixed particles have small particles taking up the pore space between the larger particles.

   D is **correct**. Size is one factor that does not affect the porosity.

10. **A** is incorrect because pumice floats because it has a low permeability. Water cannot enter the bubbles to fill them up.

    B is incorrect because a solid foundation has a very low permeability and very few pores.

    C is incorrect because loose gravel is very porous and also very permeable.

    D is **correct** because igneous bedrock has very low or no porosity. The fractures will form pathways for the water to pass through while not significantly increasing the porosity.

11. **A** is incorrect because porosity is a measure of how much empty space there is in the sediment.

    B is **correct**.

C is incorrect because permeability measures how easily water passes through a material. It does not need to be upward against gravity.

D is incorrect because saturation describes soil that is filled to capacity with water.

12. **A** is **correct**. A sea breeze depends on the Sun to heat up the land more than the water. Once the Sun stops shining, the land will cool and a sea breeze will not work.

13. **A** is **correct**. The potential for evapotranspiration (E_p) is essentially the strength of sunlight. The strength is most intense near the equator and weakens as you get closer to thee North Pole.

14. **A** is **correct**. Insolation means *in*coming *sol*ar radi*ation*.

15. **A** is incorrect because the windward side of the mountain will have a moist climate.

    B is incorrect because the top of a mountain will be cold and moist, possibly snowy.

    C is **correct** because the leeward side of a large mountain is called the rain shadow. By the time the air gets to this side of the mountain, it will be stripped of its moisture.

16. **A** is incorrect because the pore space in zone of aeration is filled with air.

    B is incorrect because aquitards have a very low permeability.

    C is **correct** because aquifers are porous and permeable.

    D is incorrect because the cone of depression is an area where water has been removed.

Motion of the Earth

Lesson 9–1: Models of the Universe

Someone comes up to you and engages you in a conversation about whether the world is moving or not. His position is that the Earth stands still and everything we experience backs up his claim (yes, there are still plenty of people around who still hold onto this belief). Your position is that the Earth spins in place every 24 hours and travels around the Sun. You get into a debate and he asks you: "How do you know the world spins?"

This is where your argument breaks down. Your only reply is that you know the Earth spins and orbits because your fourth-grade teacher told you so. But what proof is there of Earth's true motion?

Some basic observations:

- The Sun rises in the morning, travels across the sky, and sets every evening.
- The Sun rises in the eastern sky and sets in western sky.
- The Sun's path in the sky from rise to set is an arc.
- Throughout the year, the Sun is at slightly different angular altitudes in the sky at 12 o'clock noon.
- The Sun is never directly above (at zenith) in the United States (except sometimes in Hawaii).

Much to your surprise, your debate partner agrees with these observations—as it should be, because observations are observable facts and not opinions. What is being observed are apparent motions. **Apparent motion** is the way that objects *appear to move*. An apparent motion may be a real motion or a trick of perspective. Stars appear to move across the sky each night, so this would be the apparent motion of the stars. What is up for debate is whether the observed motion is real (are the stars really moving around Earth once a day?) or a trick of perspective (we are moving, and that makes the sky spin once a day).

Logically speaking, making the conclusion that the Earth sits still and the rest of the Universe moves around us is completely natural and understandable. It's a natural conclusion especially because we cannot feel any movement of the Earth. This theory is called the **Geocentric Theory,** or "Earth-centered Universe." It is very comforting to believe that the world was created just for us and everything revolves around us. It makes us feel special. It makes one feel small if you tell him that he lives on a tiny speck whizzing helplessly through the vast emptiness of space. If you are going to tell someone that he lives on a tiny speck and shatter his belief system, you need proof.

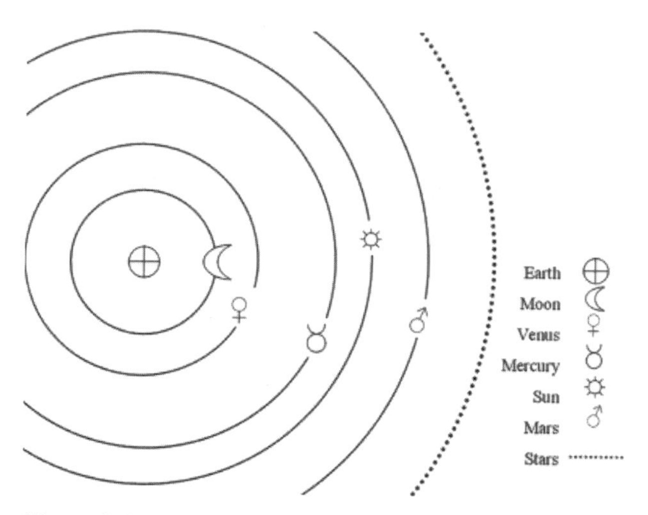

Figure 9.1. Geocentric model

The first good proof of Earth's motion came from a special pendulum constructed by Jean Foucault in 1851. The pendulum is built in a special way so that it is unaffected by the Earth's rotation. As a result, the Foucault pendulum *appears* to change its direction as the Earth, as well as us and the building, rotates on its axis.

The easiest way to describe how the Foucault pendulum works would be to imagine a pendulum set up at the North Pole. The pendulum is hanging from the ceiling of a building with a frictionless swivel. The building, which is attached to the Earth, rotates once a day while the pendulum continues to swing in its original direction. To the observer, who is also rotating along with the Earth, the pendulum appears to slowly change its direction clockwise.

The apparent change in direction of the Foucault pendulum is caused by the Coriolis Effect, which is discussed in more detail in Lesson 7–5.

The Foucault pendulum (and therefore the Coriolis Effect) is evidence of Earth's rotation. The Coriolis Effect can also be observed in the swirling patterns of large storms such as hurricanes and low pressure systems.

Galileo discovered some of the earliest evidence that proved that the Earth was not the center of the Universe or solar system, thus disproving the Geocentric Theory. Galileo did not invent the telescope, but he was the first to turn the new device to the skies and make scientific observations.

One of Galileo's groundbreaking discoveries was the discovery of four of Jupiter's moons. Galileo observed the moons orbiting around Jupiter. By definition of the Geocentric Theory, all objects revolve around the stationary Earth. By finding objects revolving around another object, Galileo disproved the Geocentric Theory.

Galileo also made some detailed observations of Venus that showed that Venus orbited around the Sun and not Earth. He saw that Venus went through phases similar to Earth's Moon. Venus also grew and shrank in size. The combination of these two observations led Galileo to the conclusion that Venus was orbiting the Sun.

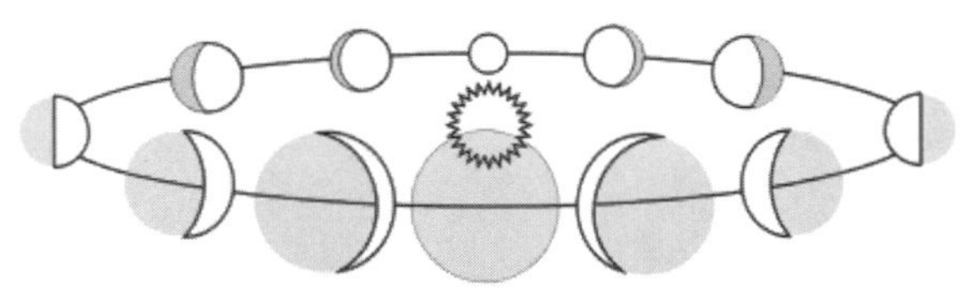

Figure 9.2. Phases of Venus

Another observation that needed explaining was that some of the planets, particularly Mars, reversed their paths across the sky for a short time—called **retrograde motion**. All planets further from the Sun than Earth demonstrated retrograde motion. The further out the planet was, the smaller the effect.

One last observation that supports the heliocentric theory (see Figure 9.3) is the changing constellations throughout the year. If you go out at midnight once a month and note the constellation directly above you, it will change from month to month. This is caused by the dark side of Earth facing different directions as Earth orbits the Sun.

In order to accommodate all of the observed phenomena and still have the Earth at the center of everything, the Geocentric Theory needed to be modified. To explain Venus's phases and changing sizes, Venus was depicted as orbiting the Sun, but the Sun (along with Venus) orbits the Earth. Jupiter's

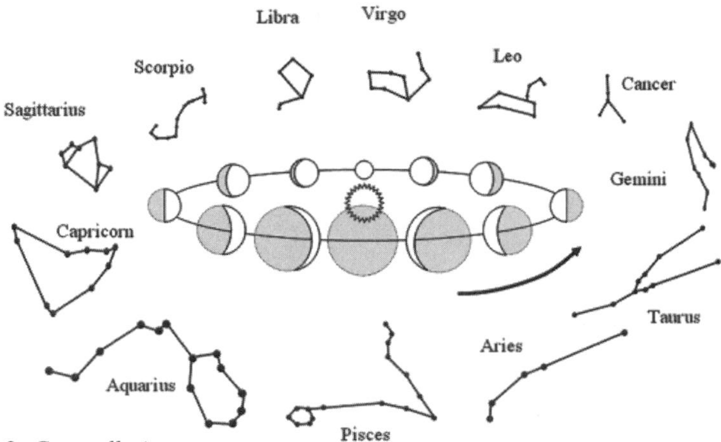

Figure 9.3. Constellations

moons orbit Jupiter and Jupiter orbits the Earth. To explain the backwards motion of the outer planets, the planets were depicted as having small orbits called epicycles, around an invisible spot and that spot orbited the Earth. With each discovery of a new planet or moon, or some detail of their motion was uncovered, an exception had to be made to the Geocentric Theory. Before long, a geocentric drawing of the solar system became incredibly complicated.

The **Heliocentric Theory** (*helio*= Sun) places the Sun at the center of the solar system. This theory accounts for all of the observations simply and without exceptions.

Lesson 9–1 Review

1. Which most accurately describes the Universe as we currently understand it?
   a) Geocentric Theory
   b) Heliocentric Theory

2. What is the name of the effect that causes the Foucault pendulum to wander around as it swings?
   a) the Foucault Effect
   b) centrifugal force
   c) the Coriolis Effect
   d) the Greenhouse Effect

3. What evidence does Venus have to disprove the geocentric theory?
   a) Venus can be seen without a telescope.
   b) We can see that Venus is curved.
   c) Venus has moons that go around it.
   d) Venus has phases that show that it orbits around the Sun.

4. What is the term for a planet appearing to move backwards in the sky?
   a) apparent motion
   b) retrograde motion
   c) geocentric motion
   d) heliocentric motion
5. What evidence did Galileo find on Jupiter that contradicts a geocentric Universe?
   a) Jupiter showed phases.
   b) He could see Jupiter spinning in place.
   c) Jupiter rises and set every day.
   d) He observed moons orbiting around Jupiter.

Lesson 9–2: Earth's Rotation

The Earth spins once a day on its axis, which is called one **rotation**. The axis is an imaginary line going from the North Pole to the South Pole through the Earth's center. If you were above the North Pole looking down at the Earth, the Earth's rotation is counterclockwise, or from west to east. It is this east-to-west rotation that makes all objects in the sky appear to move from east to west.

To put it in mathematical terms, the Earth rotates 360° in 24 hours. In each hour the Earth rotates 15° (360° ÷ 24hrs). Therefore, all objects in the sky move by 15° per hour. This is also the reason why time zones on the Earth are roughly 15° across.

A day on Earth is one complete rotation. But because our planet spins through featureless outer space, how can we tell when we have gone one complete rotation? We have a choice between using the Sun as a reference point or the distant stars. It shouldn't make much of a difference, but it actually does.

When using the stars as a reference point, it takes 23 hours and 56 minutes for the Earth to rotate 360° until the star reaches the same exact spot in the sky. This method of measuring the length of a day is called a **sidereal day**.

However, if you measure how far the Earth has to rotate until the Sun reaches the same spot, it is no longer 360°. It is closer to 361°. This is caused by the Earth's motion around the Sun. In the 23 hours and 56 minutes that the Earth was spinning, it was also orbiting around the Sun. Because the Earth has "stepped to the side" of the Sun while rotating, the

Earth has to rotate a little more to line up with the Sun. We do not have this difference with a sidereal day because the stars are essentially infinitely far away and the distance the Earth travels in one day is negligible. The extra degree of rotation takes about four minutes, bringing the total time up to 24 hours. A **solar day**, as it is called, is what we use to measure the length of a true day (sunrise to sunrise) although it does not represent a 360° turn.

Another way of tracking the rotation of the Earth is to use shadows. This is the basis for the *sundial.* You can make a crude sundial for tracking the Sun by putting a stick in the ground and observing its shadow. When the Sun rises in the east, there will be long shadow pointing west. As the Sun swings across the sky and a little south, the shadow will move eastward. Closer to noon, when the Sun gets closer to the point in the sky above you, called the **zenith**, the shadow is shortest. Then, in the afternoon, the Sun begins to head for the western horizon and the shadow gets longer and moves eastward.

> With the exception of Hawaii, there is no place in the United States where the Sun is ever directly above you (at zenith) at local noon. It is always a little to the south. Because the Sun swings across the sky to the south of zenith, the shadows always swing across the ground on the northern side of the object casting the shadow.

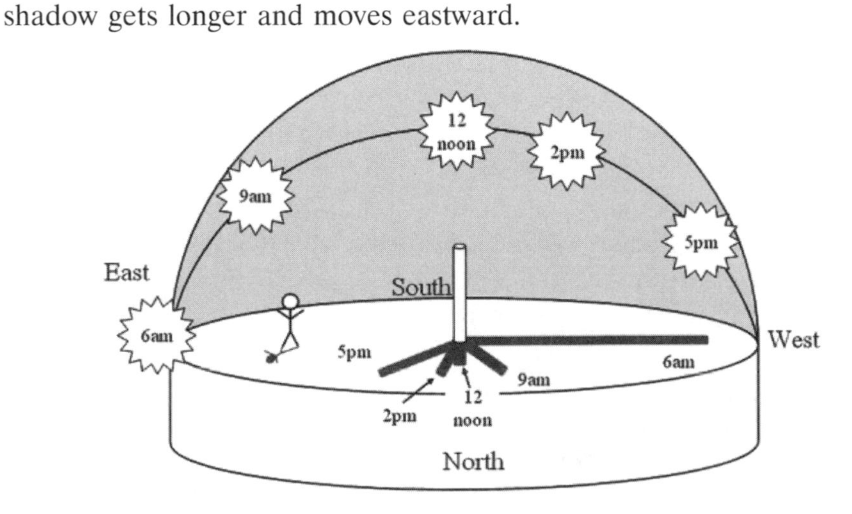

Figure 9.4. Shadows

Having Difficulty With Shadow Diagrams?

It helps if you "put yourself in the drawing." See the little person in the diagram in figure 9.4? That's you. Your shadow is always in the opposite direction as the Sun. In your mind, face the way of the drawing, picture where the Sun is, and predict where the shadow will be.

In the diagram, what time was the little person's shadow drawn?

Answer:

The shadow was drawn at 2 p.m. It is parallel to the 2 p.m. shadow from the stick.

Lesson 9–2 Review

1. How many degrees does the Earth spin in one hour?
   a) 15° c) 360°
   b) 90° d) It depends on your latitude.
2. As the Sun travels across the sky from east to west, what direction does a shadow cast by a pole move across the ground?
   a) east to west b) west to east c) south d) north
3. At noon, what general direction does a shadow cast by a pole point?
   a) north b) south c) east d) west
4. At which time(s) of the day are the lengths of shadows the longest?
   a) sunrise b) 9 a.m. c) 12 noon d) 5 p.m.
5. At which time of the day are the lengths of shadows shortest?
   a) sunrise b) 9 a.m. c) 12 noon d) 5 p.m.
6. What is the angle formed by two shadows cast one hour apart?
   a) 15° b) 45° c) 90° d) 360°
7. Is there ever a time when a vertical pole will cast no shadow because the Sun was directly overhead at noon in the continental United States?
   a) Yes, at noon on the longest day of the year.
   b) Yes, at noon on the shortest day of the year.
   c) Yes, at noon on the hottest day of the year.
   d) No. It never gets directly above.

Lesson 9–3: Earth's Revolution

The Earth's trip around the Sun, its revolution, takes 365¼ days to reach the same spot in its orbit. The Earth revolves around the Sun in 365¼ days. In order to keep our clocks running on a 24-hour system, we save up the ¼ day for four years and then add a full day to the calendar—a "leap year."

> Confused between revolution and rotation? Then memorize only one:
>
> Rotation means to spin like a drill rotates.
>
> (Revolution is "the other thing.")

Seasons

The axis of rotation is the imaginary line that the Earth spins upon. There is no real "stick" running through the Earth, and there are no real objects at the North and South Poles, but an axis is a real enough feature of any spinning object.

As the Earth rotates, it stabilizes like a spinning top. The poles point to a spot in space. The north end of the axis points to a star and as a result it is called the *North Star*. (The south end of the axis points to an empty spot in the sky so there is no "South Star.") Throughout the year, as the Earth revolves around the Sun, the axis continues to point to the North Star. This is called **Parallelism** of the Earth's axis, because our axis is always parallel to itself throughout the year.

The Earth's Axis Is Tilted!

Does this sound like another language: "The axis is tilted by 23½° from the perpendicular to the plane of its orbit"?

Because there is no "up" and "down" in space, what is it tilted from? Imagine that Earth's orbit around the Sun is a big hoop and the hoop represents the edge of a flat table. The imaginary surface of the Earth's yearly trip around the Sun is the **plane of Earth's orbit**. The axis of rotation does not stick straight up at a 90° angle from the plane of the orbit. It is tilted at 23 ½° from the perpendicular.

Seasons

To recap some of the relevant points:

- The Earth spins on its axis of rotation.
- The axis of rotation is tilted by 23½°.

- The Earth orbits the Sun once a year.
- The axis stays pointed at one spot in space all year.

The combination of these points gives us the seasons here on Earth. At one time of the year, the North Pole is tilted towards the Sun and therefore receives more solar energy. Six months later, the Earth has orbited to the other side of the Sun, and the North Pole (still pointing at the same spot in the sky) is now pointing away from the Sun. The tilting changes the angle of insolation and therefore the strength of sunlight. In fact, the Earth is closest to the Sun on January 3rd—during Winter!

Note: Nowhere in the previous paragraph is the Earth's distance from the Sun mentioned. The seasons are not caused by the distance from the Sun. Consider this: Australia has winter the same time that we have summer, yet we are both exactly the same distance away from the Sun.

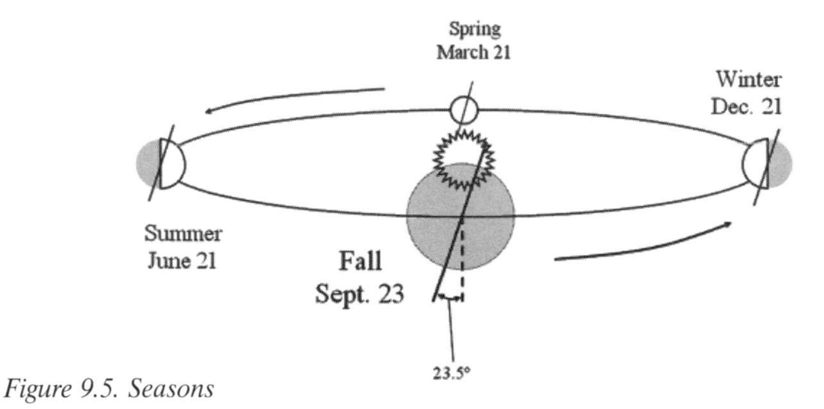

Figure 9.5. Seasons

March 21	June 21	September 23	December 21
First day of spring	First day of summer	First day of autumn	First day of winter
Vernal (spring) equinox	Summer Solstice	Autumnal equinox	Winter Solstice
Sun is over the equator	Sun is furthest north (Tropic of Cancer)	Sun is over the equator	Sun is furthest south (Tropic of Capricorn)
Day and night are equal	Longest daytime	Day and night are equal	Shortest daytime

*"Equinox" means "equal night"

Affect of Tilt on the Seasons

If the world were not tilted (in other words, the axis was perpendicular to the orbit's plane) the Sun would be located directly over the equator all year long. A situation like this would create a world with no seasons. The Northern Hemisphere would receive the same amount of energy all year. The closer to the poles you are, the less energy you receive, but it always stays at the same level. The world would perpetually be in spring.

Besides the lack of seasons, a world with no tilt would always have 12 hours of day and 12 hours of night—always at all latitudes. Right now this only happens during the equinoxes—twice a year.

If the Earth were more tilted, our seasons would go to the other extreme. Instead of no seasonal variations like a "straight up and down" world, winter and summer would get magnified. The hemisphere that is tilted towards the Sun would receive more direct sunlight for a longer period of time.

A prime example of an extreme tilt would be the planet Uranus. Uranus is tilted almost exactly on its side. During the summer in Uranus' Northern Hemisphere, the North Pole is pointed directly at the Sun—and it will be pointed in the direction of the Sun for half of its year. The North Pole is pointed at the Sun, the South Pole is in total darkness. Once the Sun sets on Uranus's South Pole, it will not be seen again for half of its year: 42 years!

Length of Day Throughout the Year

Most people in North America experience different lengths of day and night throughout the year. In winter, the days are short and in summer the days are long. Here is one way to find out how much of your 24-hour day is day and night. Use a globe and find your latitude. The circle that the latitude lines makes around the planet is also the path that you take as the planet rotates. The fraction of the circle in the daylight side represents the fraction of the day that is light. For example, ½ of the circle is 12 hours, ¾ of the circle is 18 hours.

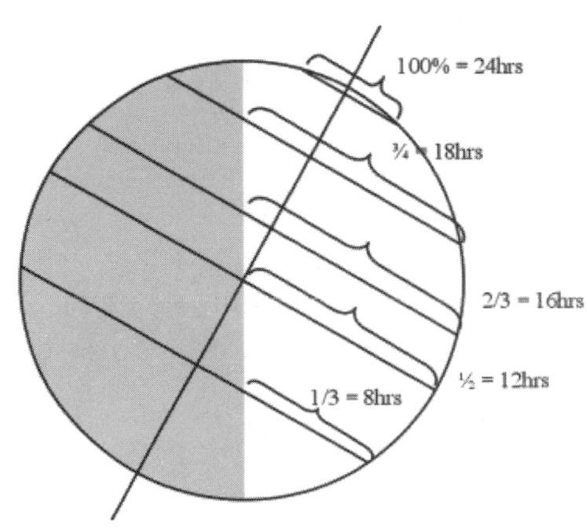

Figure 9.6. Daylight Hours

As you can see Figure 9.6, the equator always has 12 hours of day and 12 hours of night. As you get closer to the poles, the hours get further out of balance. Once you reach the Arctic Circle, the hours can get so far out of balance that it has 24 hours of day and the Sun never sets—until seasons change. The vertical line separating day and night is the **terminator**.

This is the way that the hours are distributed during summer. In winter, the pattern is reversed, but the equator still has 12 hours. Going from the summer solstice to the fall equinox, the days get shorter until they reach a balance between day and night.

Lesson 9–3 Review

1. Orbiting around the Sun is called
   a) rotation. b) revolution.
2. The Earth's axis of rotation points
   a) straight up.
   b) at the North Star
   c) in different locations as the Earth spins every day.
   d) at the equator.
3. Why do we have leap years?
   a) When the calendar was first made, an error was made measuring the length of the year.
   b) It gives an extra day off for the working class.
   c) The year is not exactly 365 days, so the extra time is saved each year until it equals a day.
   d) The Earth is spinning at a different speed than when the calendar was made.
4. The seasons are caused by the
   a) tilt of the Earth's axis. c) plane of Earth's orbit.
   b) distance to the Sun. d) greenhouse effect.
5. In winter, the North Pole receives
   a) 0 hours of daylight. c) 12 hours of daylight.
   b) 6 hours of daylight. d) 24 hours of daylight.

6. During the winter solstice
   a) the North Pole is tilted at the most toward the Sun.
   b) the North Pole is tilted at its furthest away from the Sun.
   c) the days are longest.
   d) the nights are the shortest.
7. Because there is no up and down in space, how do we determine up and down?
   a) The North Star is up.
   b) The plane of Earth's orbit is used as a reference point.
   c) Up is 90° away from the equator.
   d) Up is tilted at 23½°.
8. If the Earth's axis of rotation was greater than 23½°, what would be the effect on the seasons?
   a) Winters and summers would blend into each other.
   b) Winters and summers would be shorter.
   c) Summers and winters would be more extreme.
   d) Summers and winters would flip.
9. During winter in New York, the Earth is ________ the sun.
   a) closer to b) above c) farther from d) below

Chapter Exam

Exercise A: Seasons

Use the following diagram to do the two following activities. Note that this diagram is not exactly like the one shown earlier.

a) Draw arrows indicating the direction of the Earth's revolution.
b) For each location of the Earth, label the season and date.

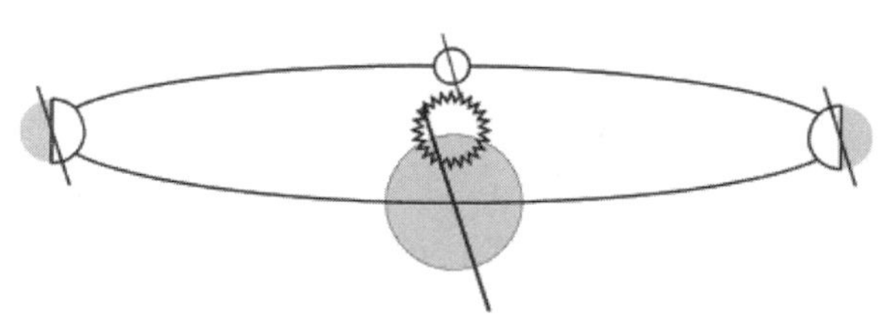

Figure 9.7

Exercise B: Day and Night

For each of the three diagrams of the Earth, perform the following:

a) Label the season and date.
b) Label the North and South Poles.
c) Using curved arrows, show the direction of rotation.
d) Lightly shade the dark side.
e) Label the number of hours of daylight for each of the following: North Pole, South Pole, Equator, latitude X.

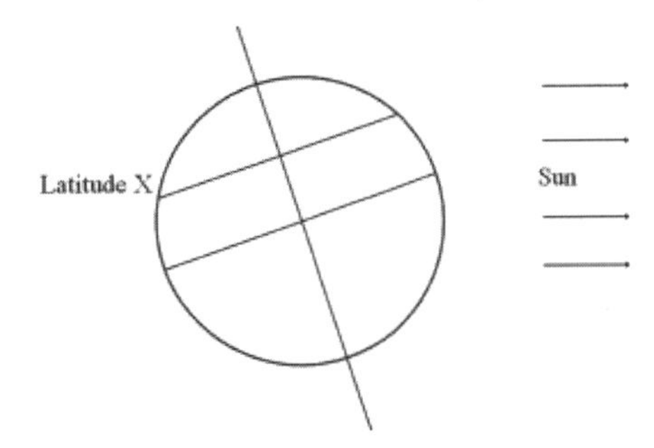

Figure 9.8. Exercise B1

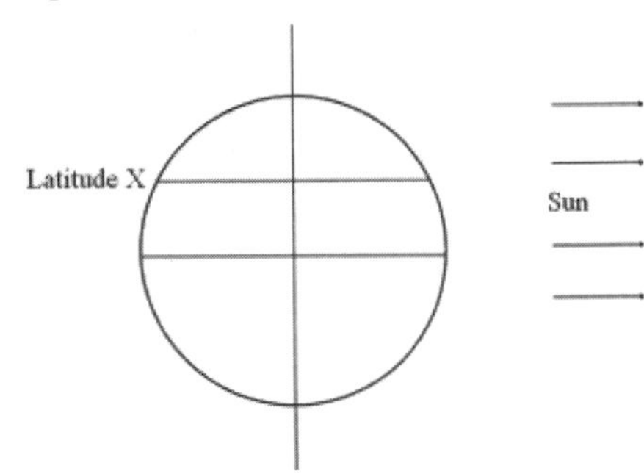

Figure 9.9. Exercise B2

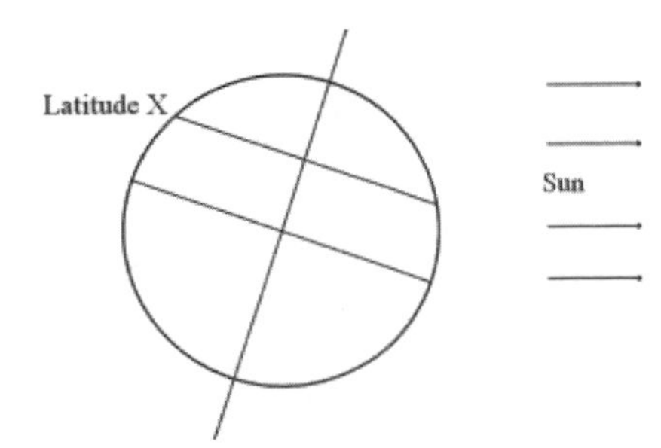

Figure 9.10. Exercise B3

Multiple Choice

1. Which theory did Galileo's observance of moons orbiting around Jupiter disprove?
   a) continental drift
   b) heliocentric
   c) geocentric
   d) relativity
2. What is the term for the appearance that a planet moves backwards because we are moving past it in space?
   a) revolution
   b) Coriolis motion
   c) transverse motion
   d) retrograde motion
3. Venus orbits around
   a) the Earth. b) the Sun. c) Jupiter. d) a large moon.
4. When you are standing at the North Pole, you will see Polaris
   a) on the horizon.
   b) at zenith.
   c) at the equator.
   d) at the same angles as your longitude.
5. You will never see the Sun directly above you at 12 noon anywhere in the United States except in
   a) Hawaii. b) Alaska. c) California. d) New York.
6. Which of the following times will give you the longest shadow on the ground in the United States?
   a) 9 a.m. b) 12 noon c) 3 p.m. d) 6 p.m.
7. One complete trip around the Sun is
   a) a rotation. b) a day. c) a solar day. d) a revolution.
8. Which of the following is the factor most responsible for the seasons on Earth?
   a) distance from the Sun
   b) the date of the equinox
   c) the angle of Earth's tilt
   d) the greenhouse effect

9. Parallelism of the Earth's axis describes
   a) the axis being parallel to the lines of longitude.
   b) the seasons beginning on the same days in parallel cultures.
   c) the Earth spinning 360° in a day which happens to be about how many days it takes to go around the Sun.
   d) the axis always pointing in the same direction in space.

10. What would the seasons be like on a world that has an axis that is perfectly vertical?
    a) The seasons will be more extreme.
    b) There will be more seasons in a year.
    c) All of the seasons will blur into one long season.
    d) It will be summer all over the planet.

Answer Key

Answers Explained Lesson 9–1

1. **B** is correct. Our current understanding is the **Heliocentric Theory**, which puts the Sun at the center of the solar system (although not the center of the Universe).

2. **A** is incorrect because there is no such thing as a Foucault Effect.

   B is incorrect because centrifugal force is the force that pins you to the car door as you make a fast turn.

   C is **correct**.

   D is incorrect because the greenhouse effect causes the Earth to warm up from trapped heat.

3. **A** is incorrect because that can be explained in both theories.

   B is incorrect because it doesn't prove that the Earth is or isn't at the center of the Universe.

   C is incorrect because Venus does not have moons. The moons were seen orbiting around Jupiter.

   D is **correct** because the phases demonstrate that Venus changes angles as it moves around the Sun.

4. **A** is incorrect because apparent motion is any motion that we see caused by either our motion or the other object's motion.

   B is **correct**. *Retro* means backwards.

   C is incorrect because geocentric motion means Earth-centered motion.

   D is incorrect because heliocentric motion means Sun-centered motion.

5. **A** is incorrect because Jupiter does not show phases.

   B is incorrect because Galileo was not able to see Jupiter spinning.

   C is incorrect because Jupiter rising and setting does not contradict the Geocentric Theory.

   D is **correct** because moons orbiting around Jupiter show that objects orbit around something besides the Earth.

Answers Explained Lesson 9–2

1. **A** is **correct**. 360° ÷ 24hrs = 15°

   B and **C** are incorrect.

   D is incorrect because the entire Earth spins at an even 15° every hour—the equator and the poles included.

2. **A** is incorrect because the shadows move in the opposite direction than the Sun.

   B is **correct** because shadows move in the opposite direction to the Sun's motion.

   C is incorrect. In the United States, the Sun is always to the south so the shadow swings on the northern side of the object casting the shadow.

   D is incorrect. The shadow will swing to the northern side of an object but it will travel from west to east.

3. **A** is **correct** because the Sun is always south of us at 12 noon.

   B is incorrect. Because the Sun is to the south at noon, the shadow will be cast in the opposite direction.

   C and **D** are incorrect.

4. **A** is **correct**. When the Sun rises (and also when it sets), it casts a very long shadow towards the west.

   C is incorrect because at noon, the shadows are shortest.

   B and **D** are incorrect.

5. **A** is incorrect because at sunrise, the Sun is on the horizon, which gives a very long shadow.

   C is **correct** because at noon the Sun is at its highest point in the sky and closest to being directly above you.

   B and **D** are incorrect.

6. **A** is **correct**. The Earth spins 15° per hour, which makes the Sun appear to move across the sky at 15° per hour. Any shadows that the Sun makes will shift at 15° per hour.

   B is incorrect because 45° is how far the shadow will move in three hours.

   C is incorrect because 90° is how far the shadow will move in six hours.

   D is incorrect because 360° is how far the shadow will move in 24 hours.

7. **D** is **correct** because, in the continental United states, the Sun never reaches zenith.

Answers Explained Lesson 9–3

1. **B is correct.** Orbiting around the Sun is called **revolution**. Rotation is to spin on an axis.
2. **A** is incorrect because there is no "straight up" in space. Also, the axis points 23½° away from "straight up."

   B is **correct**. The axis of rotation is lined up with the North Star, which is why it is so special.

   C is incorrect because the axis of rotation is very stable and will continue pointing at the same spot for hundreds of years.

   D is incorrect because the equator is 90° away from the axis.
3. **C** is **correct**. The Earth takes 365¼ days to orbit the Sun. The ¼ day is saved for three years and then added into the calendar.
4. **A** is **correct**. The tilt of the planet changes the angle of insolation throughout the year.

   B is incorrect because when it is winter in one place, it is summer in another. Both places are at the same distance from the Sun.

   C is incorrect because it is the Earth's tilt with respect to the plane of orbit that gives us seasons.

   D is incorrect because the greenhouse effect will adjust the seasons that are already here.
5. **A** is **correct** because the North Pole is tilted away from the Sun and will not rotate onto the day side of Earth for months.

   B, **C**, and **D** are incorrect.
6. **A** is incorrect because the North Pole is tilted away from the Sun on the Winter Solstice.

   B is correct.

   C and **D** are incorrect because the winter solstice is the day of the shortest day and longest night.
7. **A** is incorrect because The axis of rotation points to the North Star and the axis is tilted by 23½°.

   B is **correct**. North for our planet is 23½° away from being perpendicular to the plane of revolution.

   C is incorrect because 90° away from the equator is the North Pole and Polaris, both of which are 23½° away from "up."

   D is incorrect because the axis of rotation is tilted at 23½°.

8. **A** is incorrect because the seasons will blend on a world where the axis is straight up and down (no tilt).

   B is incorrect because the seasons would be the same length of time but there would be a quicker transition from one to the next.

   C is **correct** because, when the North Pole is tilted towards the Sun, it would be tilted more giving more heat to the Northern Hemisphere for summer.

   D is incorrect because the seasons would still happen in the same order.

9. **A** is **correct**. The seasons are caused by the tilt of the axis. During winter, the North Pole is tilted away from the Sun.

Answers Explained Chapter Exam

Exercise A: Seasons

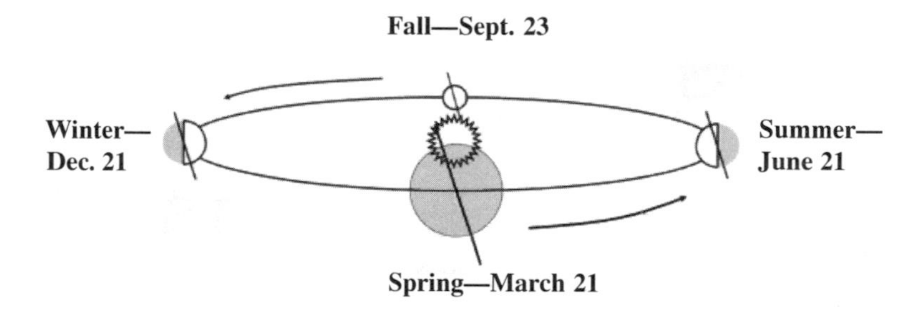

Exercise B Day and Night

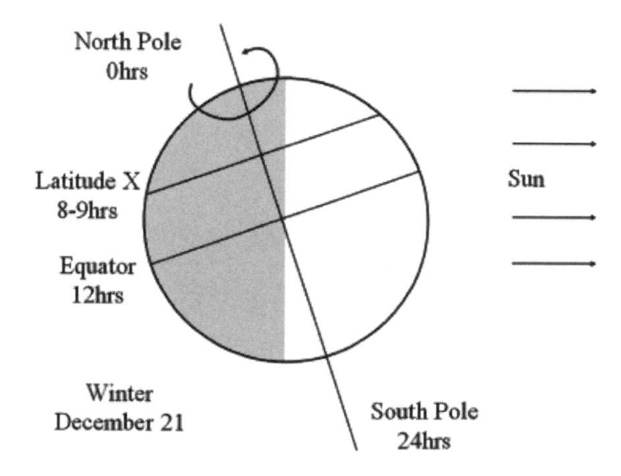

Figure 9.12. Exercise B1 Answer

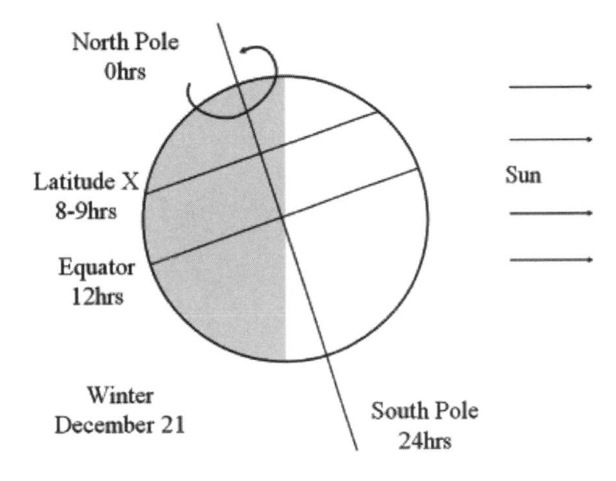

Figure 9.13. Exercise B2 Answer

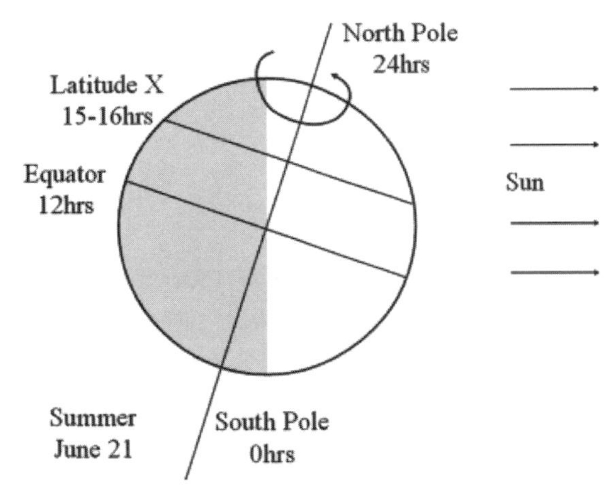

Figure 9.14. Exercise B3 Answer

Multiple Choice

1. **C** is **correct** because if there are objects in space that orbit something besides the Earth, then the Earth cannot be the center of the Universe.
2. **A** is incorrect because revolution means to orbit around and object.

   B is incorrect because Coriolis motion turns to the right in the Northern Hemisphere.

   C is incorrect because transverse motion is what two plates do when they slide past each other.

   D is **correct**.

3. **B** is **correct**.
4. **D** is **correct** because the North Pole points at Polaris and the zenith is the spot in the sky directly above you.
5. **A** is **correct** because Hawaii is less than 23½° above the equator.
6. **A** and **C** are incorrect.

   B is incorrect because 12 noon will give you the shortest shadow.

   D is **correct** because at 6 p.m., the sun is lowest in the sky.
7. **A** and **B** are incorrect because a rotation is a spin on the axis of rotation. It is a day.

   C is incorrect because a solar day is a spin around on the axis of rotation until the sun returns to the same exact position.

   D is **correct** because a revolution is an orbit or trip around the sun.
8. **A** is incorrect because the distance is the same for the United States and Australia, yet they have summer and winter at the same time.

   B and **D** are incorrect.

   C is **correct**.
9. **A** is incorrect because all lines of longitude are not parallel, they intersect the poles.

   B and **C** are incorrect.

   D is **correct**.
10. **A** is incorrect because a more tilted axis would make more extreme seasons.

    B is incorrect because a vertical axis would make only one season for the whole year.

    C is **correct** because each location on the planet would be heated equally all year.

    D is incorrect.

Astronomy

Lesson 10–1: Ellipses

Johannes Keppler was the first astronomer to accurately describe the motion of the planets around the Sun. His descriptions fell into three sets of relationships, which are now called **Keppler's Laws of Planetary Motion**.

Before I discuss Keppler's Laws, we need to have a working knowledge of ellipses. A circle is type of ellipse. One way to draw a circle is to use a loop of string, a pin, and a pencil. To draw an ellipse, use a loop of string and *two* pins with a little separation between them. This will essentially draw an oval. Ellipses can range in shape from a perfect circle (a circle is a special type of an ellipse) to an oval that is stretched so tight that it becomes a flat line.

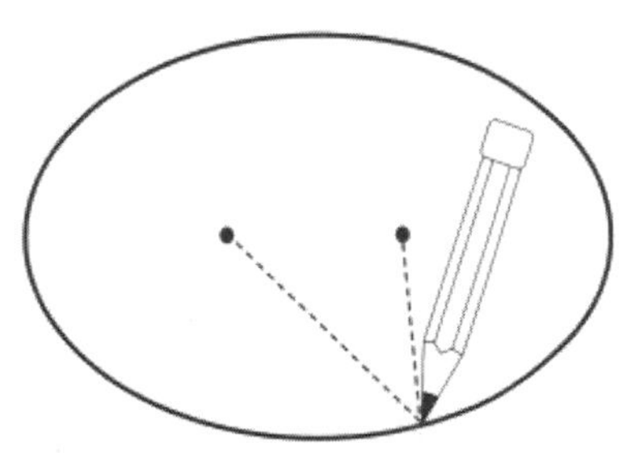

Figure 10.1. Drawing an ellipse

In the case of the ellipse, the two pins are each called a **focus**.

One definition of a **circle** is: "All the locations that have the same distance from a central point."

The definition of an **ellipse** is: all the locations that have the same *added* distances from two foci ("focuses").

This definition means that the distance from pin A to a spot on the ellipse, plus the distance from pin B to the same spot, will always be the same value. A **focus** is a center point of an ellipse.

The **eccentricity** of an ellipse is a measure of how "weird" a circle it is. A perfect circle is not very weird at all and therefore has an eccentricity of zero. A very elongated oval is a "very weird circle" and therefore has a big eccentricity. Eccentricity has a range from zero to one with zero being a perfect circle and one being a flattened ellipse or a line.

To calculate eccentricity, use the following equation:

$$e = \frac{d}{L}$$

where d is the distance between the two foci and L is the length or width of the ellipse.

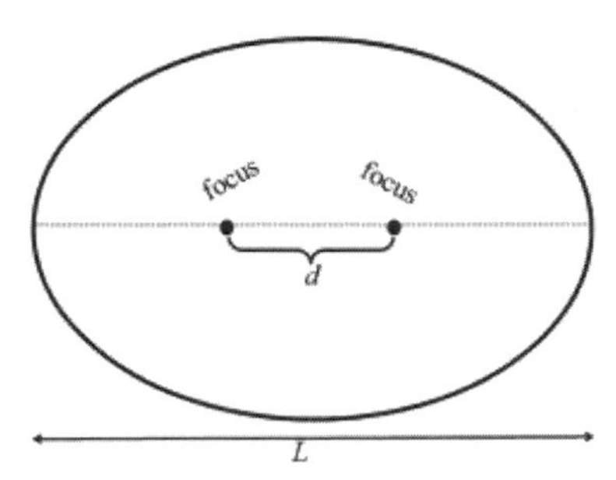

Figure 10.2. Anatomy of an ellipse

If *d* is zero, which means that you have both foci in one spot, then e becomes zero—a perfect circle. If *d* gets bigger, the two foci get spread apart and you get a flattened oval.

Keppler's First Law

All planets travel in elliptical orbits with the Sun at one focus.

All planets' orbits are ellipses. Most are so close to a perfect circle that we can't see that they are ovals. Pluto and some of the minor planets (asteroids) have highly eccentric orbits.

A typical question at this point is: "If the Sun is at one focus, then what's at the other?" The disappointing answer to that question is that there is nothing special at the second focus' position in space. The second focus is a tool that we use on paper in order to allow us to accurately draw the shape of planets' orbits.

Keppler's Second Law

Each planet travels in such a way that a line joining the planet and the Sun sweeps equal areas in equal times.

This law is best understood by looking at a diagram. (See Figure 10.3.)

Each location on the diagram is separated by one month of time. Notice that in the same amount of time (a month) the planet travels different distances, which means that the planet is traveling at different speeds. Also notice that the planet moves faster (bigger distance each month) as it gets closer to the Sun. Due to the elliptical orbit in which all planets travel, there will be times of the year that the planet is closer to the Sun than others.

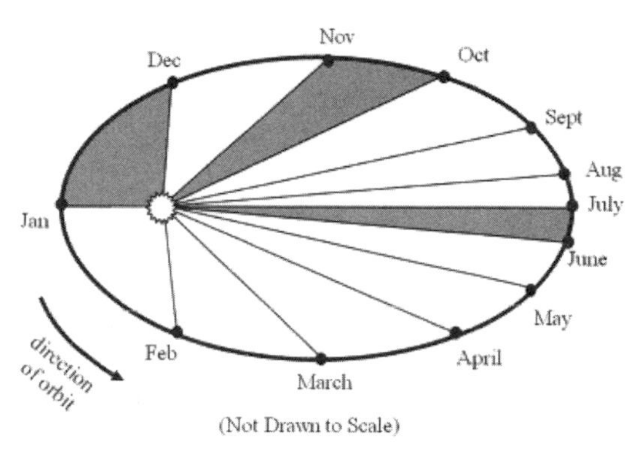

Figure 10.3. Equal areas

To go back to the wording of the law stated previously, let's discuss the meaning of "equal areas in equal times." Keppler imagined a long rubber band attached from the Sun to the Earth. As the Earth travels in its orbit, the rubber band sweeps through a section of space. In the span of a month, the band will sweep through an area of space shaped like a pizza pie slice. Each month will be a new slice of pizza but because the planet is constantly changing speed and distance the shape of the slice will differ.

Despite the different shapes of slices, Keppler realized that each slice of the orbit will have the same area after the same time. The time interval does not have to be a month; it could be a week or a day. As long as the time intervals are the same, the areas swept by an orbiting planet will be the same. For this reason, this law is sometimes called the "Law of Equal Areas."

Keppler's Third Law

The square of the planet's period is proportional to its distance from the Sun, cubed.

The further a planet is from the Sun, the longer it takes to go around the Sun. (translated into real English)

This relationship is best expressed as an equation:

$$P^2 = D^3$$

where P is the amount of time it takes for a trip around the Sun. This is called a "period" in Earth Years.

And D is the distance in "AUs."

An "AU" is an **astronomical unit**, which is the average distance from the Sun to the Earth (approximately 93,000,000 miles or 150,000,000 km).

Of course it takes longer for a planet to orbit if it is farther away, but there are a couple of reasons why this happens. The most obvious reason is the planet's longer path around the Sun. The other reason for the longer period is the slower speed of the planet. If a planet is further from the Sun, the Sun's gravity will have a weaker pull the planet. If the planet moves too fast in the weak gravitational pull, it will escape from the Sun.

Sometimes Keppler's Third Law is called "The Law of Harmonic Motion" because the planets orbit in harmony with each other, faster in towards the Sun and slower further out. In all, the bodies in the solar system move in a graceful, poetic fashion.

The same relationships apply to objects other than planets—such as satellites. A natural satellite is any object that orbits a planet. Satellites orbiting a planet sweep equal areas in equal times and orbit slower when they are further. Even comets, with their highly elongated orbits follow Keppler's laws. When they get close to the Sun, they move very fast and continually slow down as they reach the far ends of their orbits.

Lesson 10–1 Review

1. Which of the following ellipses is most eccentric?

a)

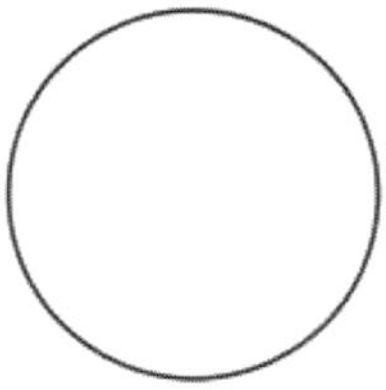

Figure 10.4 Ellipse A

c)

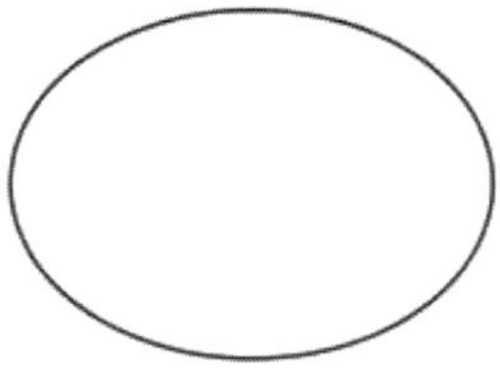

Figure 10.6 Ellipse C

b)

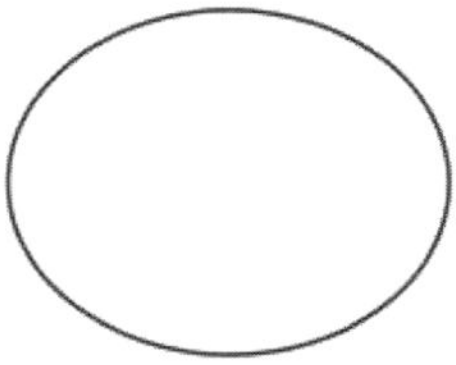

Figure 10.5 Ellipse B

d)

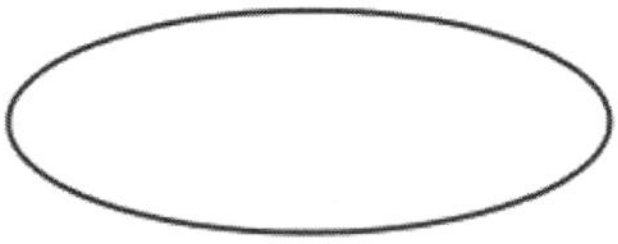

Figure 10.7 Ellipse D

2. Which of the previous ellipses represents the true shape of Earth's orbit?

a) A b) B c) C d) D

3. What is the eccentricity of an ellipse where the focal distance is 4 cm and the major axis is 10 cm?
   a) 2.5 b) .25 c) .4 d) 4.0

4. What effect does increasing the distance between the foci have on the shape of the ellipse?
   a) Eccentricity increases.
   b) Eccentricity decreases.
   c) Eccentricity stays the same.
   d) Eccentricity does not change.

5. What is the maximum eccentricity an ellipse can have?
   a) .99 b) 0 c) .25 d) 1.0

6. Where would the location of the Sun be on a scale drawing of the Earth's orbit?
   a) in the exact center
   b) at one of the two foci
   c) outside of the orbit
   d) at both foci

7. What is the minimum eccentricity an ellipse can have?
   a) 1.0, which forms a perfect circle.
   b) 1.0, which forms a straight line.
   c) 0.0, which forms a perfect circle.
   d) 0.0, which forms a straight line.

8. The Earth's orbit looks like
   a) a circle. b) an ellipse.

Lesson 10–2: The Moon

The Moon is the Earth's only natural satellite. There are several theories as to how the Moon formed but they fall into two main categories: It was formed at the same time as Earth out of the gas and dust that formed all of the solar system, or it was created when a piece of Earth was thrown off into space.

One formation theory states that the solar system (the Sun, planets, and everything else in it) was formed from a huge cloud of gas and dust in space. The gas and dust contracted by gravity and coalesced into all the objects that are around now. Some of the planets formed with smaller clumps orbiting around them, which eventually contracted into natural satellites—moons. This theory makes sense because most satellites orbit their planet very close to the plane of the planet's equator, as does our Moon. The vast majority of moons orbit in the same direction that the planets orbit as well as spin (counterclockwise as seen from above north).

The other leading theory states that the Earth formed first as the cloud of dust and gas contracted. At some time after formation, the Earth was struck by a very large object (about the size of Mars). The collision threw out a huge chunk of the Earth. This chunk reformed out in space and began orbiting the Earth.

The Moon revolves around the Earth once a month which is probably where the word *month* ("moonth"?) came from. The Moon revolves 360° around the Earth once every 27 1/3 ! days in a period called the **sidereal month**. However, it takes closer to 30 days from one full Moon to the next, which is called a **synodic month,** or a lunar month. This happens because the Earth continues to travel around the Sun, and the Moon needs to catch up.

As previously mentioned, the Moon takes 27 1/3 days for one revolution of 360° around the Earth. That means that the Moon orbits about 13° each day. If you go outside every day at exactly the same time, the Moon will be 13° further to the east than it was the day before.

$$360° \div 27\frac{1}{3}\text{ days} = 13°/\text{day}$$

Therefore, it takes the Earth an extra 50 minutes to turn far enough for the Moon to be in the same exact spot in the sky as it was the previous day. The most obvious time to make this observation is at moonrise. The Moon will rise 50 minutes later each night. If you observe this for 27 nights in a row, the Moon will be back where it started a lunar month ago.

Phases of the Moon

As the Moon revolves around Earth, the angle between the Earth, Moon, and Sun constantly change. The changing angles cause the Moon to go through phases. The side of the Moon that faces the Sun is always lit up. However, from our place here on Earth, we will see different portions of the lit side on the Moon.

The phases are traditionally listed starting with the New Moon. Following the new moon—which cannot be seen from Earth—the phases get "larger" until the Moon is full. This is called the **waxing** portion of the phases. After the full moon, the phases begin to get "smaller" until the next New Moon. This is called the **waning portion of the phases**.

To Help You Remember. . .

Waxing means to get brighter. When janitors wax the floor at school it gets shiny.

Waning means to get dull. After the floors are walked-on, the shine wanes.

There is a discrepancy with the Half Moon phase. Some sources list it as a Half Moon; others call it a Quarter Moon. It is called a Quarter Moon because the Moon is one quarter of the way through its cycle. Luckily, there is no other phase that is called a quarter or a half Moon so there's *no confusion*.

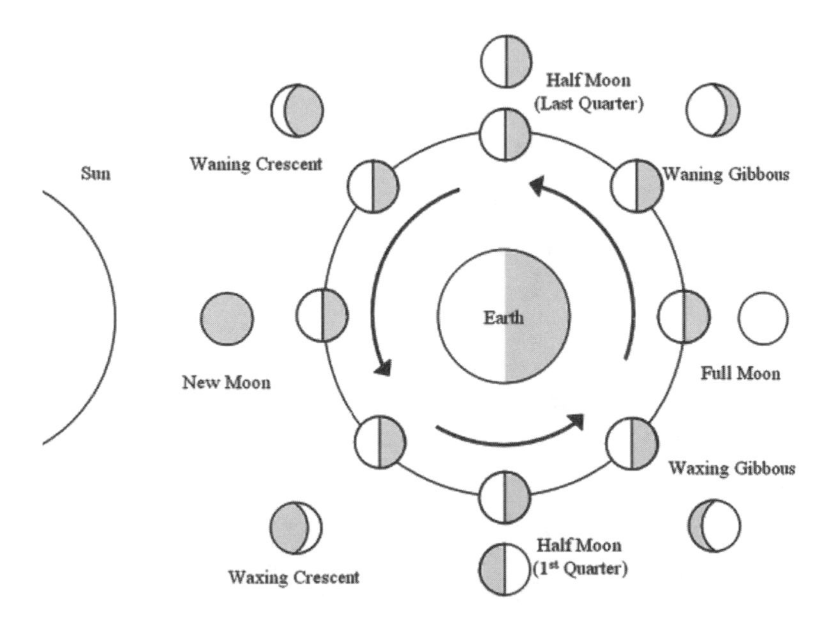

Figure 10.8. In this figure, the outer circle of moons represents what we see from Earth.

Gibbous means "egg-shaped."

One popular misconception is that the Moon is only visible during the night. In fact, the Moon is visible during the day just as much as at night with the exception of the New Moon. The Moon is not noticed as much during the day because most often it will be in a smaller phase hidden in the bright daytime sky. It can be easily seen if you know where to look.

Lesson 10–2 Review

1. What does the term *waxing* mean with respect to the phases of the moon?
   a) They are the phases that are only visible at night.
   b) They are the phases while the Moon is getting brighter.
   c) They are the phases that are only visible during the daytime.
   d) They are the phases while the Moon is getting dimmer.
2. How often is the Moon visible during the day?
   a) never
   b) every day that it is clear
   c) about half of the month
   d) two days a month
3. If the Moon rises at exactly 8:00 tonight, what time will it rise tomorrow?
   a) 7:10 p.m. b) 8:00 p.m. c) 8:50 p.m. d) 8:10 p.m.
4. Why can't the Moon be seen during a new moon?
   a) The Moon is on the day side of the Earth and can't be seen during the day.
   b) You can see it if you look in the right place.
   c) It is being eclipsed.
   d) We are looking at the darkened side of the Moon very close to the Sun and through the bright blue sky.
5. Which category best classifies the Moon?
   a) asteroid b) planet c) satellite d) meteor
6. Which phase follows the waning crescent?
   a) new moon
   b) 1st quarter
   c) waning gibbous
   d) last quarter

Lesson 10-3: Eclipses

The word *eclipse* can be loosely translated to mean "gets dark." Here, on Earth, we have two types of eclipses: a lunar eclipse ("Moon gets dark") and a Solar Eclipse ("Sun gets dark"). Either eclipse will happen whenever the Earth, Sun, and Moon are lined up in a straight line.

Theoretically, everything should line up twice a month: once during the new moon and once during the full moon. However, we don't have an eclipse twice a month or even once a month. This has to do with the tilt of the Moon's orbit. Imagine that the moon's orbit is a hoola-hoop around Earth and the hoop has a slight tilt. Most months, the tilt causes the Moon's shadow to miss the Earth during a new moon, and the Earth's shadow to miss the Moon during a full moon. Only when the tilt causes the Moon to be directly in line with the Sun and Earth is there an eclipse.

Solar Eclipse

A **solar eclipse** is one of the most dramatic celestial events that can be witnessed. During the typical new moon, the Moon is between the Earth and Sun but not exactly in a straight line. When this happens, the lit side of the Moon faces away from us and we look at the darker side against the bright sky.We cannot see the moon. But, when the Moon is lined up exactly between Earth and Sun, we see the Sun's light get blocked. This is a **solar eclipse**.

The most common type of solar eclipse is the partial eclipse. A **partial eclipse** happens when the Moon is just off of being directly between Earth and the Sun. During this eclipse the Sun will appear as a cookie with a bite out of it. These eclipses are rated by the percent of the Sun that is blocked.

Even though the Sun is very large compared to the Moon, the Moon can block the entire Sun because the Moon is so much closer to Earth. In a similar fashion, you can "eclipse" a friend you are looking at by holding the hand close enough to your eyes. The Moon has the perfect combination of size and distance to completely block out the disc of the Sun.

The Earth/Moon system is the only one in the solar system where a planet's moon is just the right size and distance in order to make a total solar eclipse.

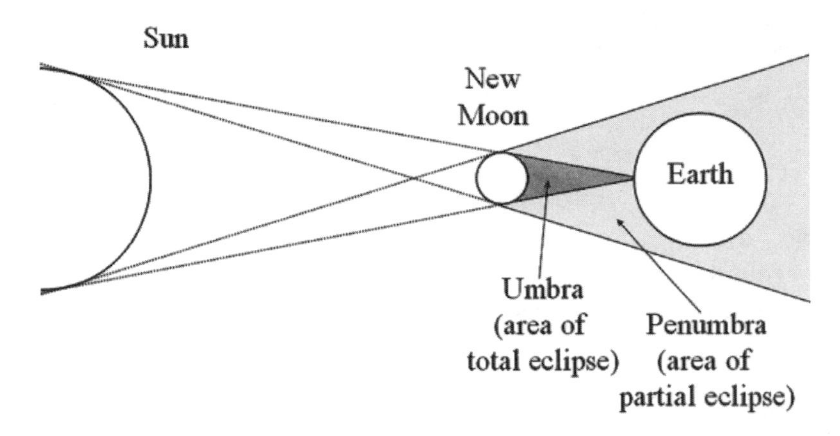

Figure 10.9. Solar eclipse

The Moon obeys Keppler's laws, and it orbits the Earth in an elliptical orbit. Therefore, the Moon's distance from Earth changes slightly. When the Moon is a little closer, it will appear larger in the sky and can cover the Sun with a little room to spare. When the Moon is a little further from Earth, it appears smaller and may be too small to completely block the Sun. When this happens, there will be a ring of the Sun that will be visible at the peak of the eclipse. This type of eclipse is called an **annular eclipse**.

From space, a solar eclipse that is happening on Earth will look like a bull's-eye. Projected on the Earth's surface will be a small, dark shadow surrounded by a larger, less-dark shadow (the Moon creeping across the Earth). The only the people on Earth who will see a total eclipse are those within the dark shadow, the **umbra**. Outside of the darkest shadow and within the dimmer shadow, the **penumbra**, observers will see a partial eclipse. Less and less of the Sun will be blocked as the observers get further away from the center of the shadow. Outside of the shadows, observers will experience a new moon. From start to finish, a solar eclipse will last for a couple of hours with the period of *totality* (the phase of the eclipse during which it is total) being as much as 7 minutes.

Why Is it Dangerous to Look at a Solar Eclipse?

As anyone who has looked directly at the Sun knows, it is quite uncomfortable. Your eyes water and you see spots. The spots you see are actually burns on the back of your eyes (the retina). At first, the burns are minor and will go away quickly, but they get worse as you keep looking.

During an eclipse, the Sun is still incredibly bright and will continue to damage your eyes as you look more and more. In addition, the disc of the Sun is not the only part that gives off harmful light. The Sun's "atmosphere," the **corona**, gives off light, but it is usually too dim to see next to the bright Sun. The corona does not get covered up by the Moon during an eclipse. Besides visible light, the Sun also gives off ultraviolet (UV) light, which can also cause damage to your eyes.

You should NEVER use a camera, telescope, or binoculars to directly view a solar eclipse. This will instantly cause permanent damage to your eyes. There are special filters for telescopes and binoculars that can be used to safely view the event.

What's a safe way to view it?

A telescope can safely be used to project the image onto a wall or board as long as you don't look through the scope (this technique can damage some sensitive optics is some telescopes). Sunglasses are not dark enough. A welding visor can be used.

One technique that is often sited is the pinhole camera. Take a piece of thin cardboard and poke a small hole through it. Hold the cardboard out to your side and let the Sun's light pass through it and fall on the ground. It will cast an image showing the eclipse as it progresses.

A variation of the pinhole camera that I discovered during a partial eclipse that gave a striking result was to use my kitchen colander (a bowl with holes in it for straining water from food). The many holes in the colander cast dozens of little eclipse-shaped images on the floor.

Lunar Eclipse

A **lunar eclipse** happens when the Moon crosses into the shadow cast by the Earth during a full moon. During a "normal" full moon, the Sun, Earth, and Moon will be in an almost-straight line and the Moon will not cross into Earth's shadow because if the tilt of the Moon's orbit. During a lunar eclipse, the Moon will drift through Earth's shadow and get dimmer.

Because the Earth is so much larger than the Moon, the entire Moon can be inside Earth's shadow with room to spare. For this reason, lunar eclipses happen more often than solar eclipses. In addition, the darkened Moon is visible to everyone on the entire nighttime side of the Earth, rather than people in a small area as with a solar eclipse.

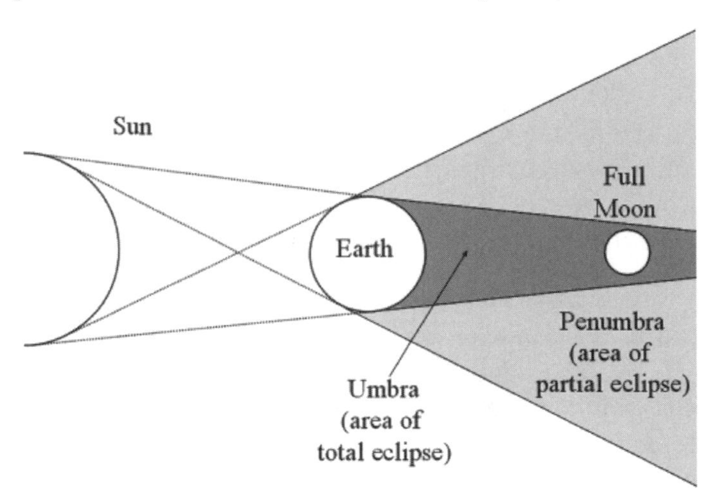

Figure 10.10. Lunar eclipse

A lunar eclipse can be a partial eclipse or a total eclipse just as its solar counterpart can. But there is one major distinction between a solar and lunar eclipse: a total lunar eclipse does not blacken out—it makes the Moon dimmer. This is caused by refraction of sunlight through Earth's atmosphere. Most of the light gets blocked behind the Earth, but some sneaks around the edges through our atmosphere. The light skimming through the atmosphere is essentially "sunset light," so it is usually reddish in color. During a total lunar eclipse, the Moon will take on a deep blood red or bronze color at its peak. It's no wonder that many cultures believe that a lunar eclipse is an omen of death and war!

Lesson 10–3 Review

1. When there is a lunar eclipse, what phase will the Moon be in?
   a) new moon
   b) waxing moon
   c) full moon
   d) waning moon

2. Why don't we have a solar eclipse every month?
   a) The moon's orbit is tilted.
   b) The Earth's axis is tilted.
   c) The tides throw off the alignment.
   d) The Moon is usually too far away.
3. If you were standing in the penumbra during a solar eclipse, what would you observe?
   a) a total eclipse
   b) a partial eclipse
   c) a regular new moon
   d) sunset
4. If a solar eclipse happens while the Moon is at its furthest point in its elliptical orbit, what will you experience?
   a) a partial eclipse
   b) a total eclipse
   c) an annular eclipse
   d) a lunar eclipse
5. Lunar Eclipses are more common than solar eclipses because
   a) the Earth's shadow on the Moon is larger than the moon's shadow on the Earth.
   b) the Earth's orbit is tilted less than the moon's.
   c) new moons are more rare than full moons.
   d) because of Keppler's laws, the Moon moves faster during the new moon phase.

Lesson 10–4: Tides

The changing tides of the oceans are caused by the gravitational forces between the Earth, Sun, and Moon. The Universal Law of Gravitation tells us how strong the pull will be between two objects near each other.

The Universal Law of Gravitation

$$F \propto \frac{m_1 m_2}{d^2}$$

Where F is the force of gravity,
m_1 and m_2 are the masses of the two objects involved, and
d^2 is the distance between the objects squared.

The funny $\propto$ symbol stands for "is proportional to" and essentially means "equal to."

This equation tells us that the gravitational force between two objects gets stronger as the masses get larger and that the force gets weaker as the objects get farther away from each other. The Moon is the closest object to Earth and has the strongest influence. The Sun is the heaviest object around, but is fairly far away, and it has the second strongest pull on the Earth.

Do the Other Planets Pull On the Earth?

Technically speaking, the Earth is affected by the gravitational pull of every object in the Universe. So, yes, the other planets do pull on the Earth. However, with the exception of the Sun and Moon, the gravitational pulls from other objects are insignificant.

If the Planets Line Up, Will it Pull the Earth Apart?

No. In the billions of years that the solar system has existed, the planets have undoubtedly lined up, and yet the Earth is still here. In fact, you have a stronger gravitational pull on the Earth than the next nearest object (except maybe the Moon and Sun).

The Causes of the Tides

To oversimplify the causes of the tides, think of it this way: The Moon's gravity pulls on the Earth and pulls strongest on the spot of the Earth that is directly under the Moon. The solid portion of the Earth does not get pulled up to the Moon because the solid rock is firmly attached. The liquid oceans are free to move and get pulled up to the Moon just a little bit, making a small hill of water. This hill of water stays at the spot under the Moon as the Earth rotates once a day under it—making the hill appear to migrate around the Earth. Wherever the hill is located, there will be a higher tide.

This oversimplification does not account for the fact that most places on Earth have TWO high tides and TWO low tides each day. Besides having a high tide directly under the Moon's position, there is also a high tide on the opposite side of the Earth. This bulge of water is actually being "thrown off" of thc Earth due to centripetal force (the force that causes you to slide against the car door during a tight turn).

Imagine two young girls playing in the park. They grasp each others' hands and twirl each other around while facing each other. As they spin faster and faster, they have to lean away from each other and their long hair gets thrown outward from the rotation.

The Earth and Moon do a similar dance with each other. They each spin around a common balance point called the **barycenter**. Because the Earth is so much more massive than the Moon, the barycenter is much closer to the Earth. This is much like a parent and child on a seesaw. The parent has to adjust the seesaw so that he is closer to the fulcrum and the child's end is extended far from the center. The barycenter is located approximately 1,700 km (1,060 mi) below Earth's surface and both objects orbit around this common point. As a result, the Moon makes large orbiting circles and the Earth wobbles around the point.

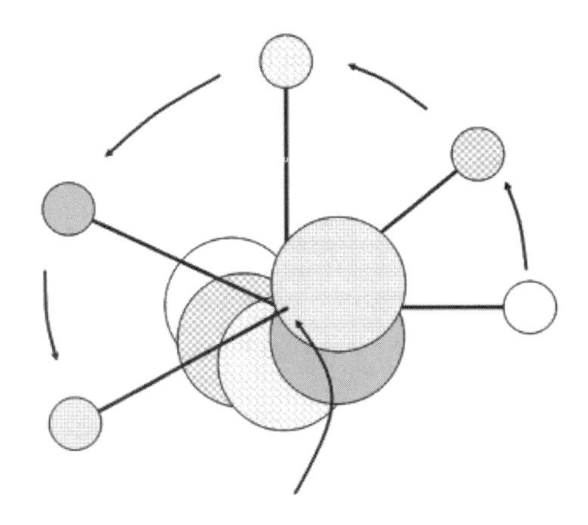

Figure 10.11. Barycenter

As the Earth wobbles around the balance point between the Earth and Moon, the water on the far side of the Earth gets thrown out into space a little. Sort of like a liquid being swirled in a glass. This is the cause of the second high tide bulge.

Even though the Moon has the biggest gravitational influence on the Earth and its oceans, we also need to consider the affects of the Sun's gravity, which also tugs on the ocean water, but to a lesser degree than the Moon's gravity. When the Sun and Moon are lined up, they work as a team and pull extra hard on the water. The result: an extra high tide called a **spring tide**. When the Sun and Moon are 90° away from each other in our sky, they pull the water in two completely different directions and essentially "share" the bulge of water. This makes the high tide lower than it normally can be. This is called a **neap tide**.

A **spring tide** happens whenever the Earth, Moon, and Sun are in a straight line—or in other words: during a full or new moon. A spring tide gets its name from the springing action of the water as it goes from its highest possible tide to its lowest possible for the month. It springs up and down. A spring tide can happen every month and season, not just during the season that follows winter!

The **neap tide** will also happen twice a lunar month when the Sun, Earth, and Moon make a 90° angle. This happens during both of the half-Moon phases (also called the first and third quarters). A neap tide will produce high tides that are not very high and low tides that are not very low.

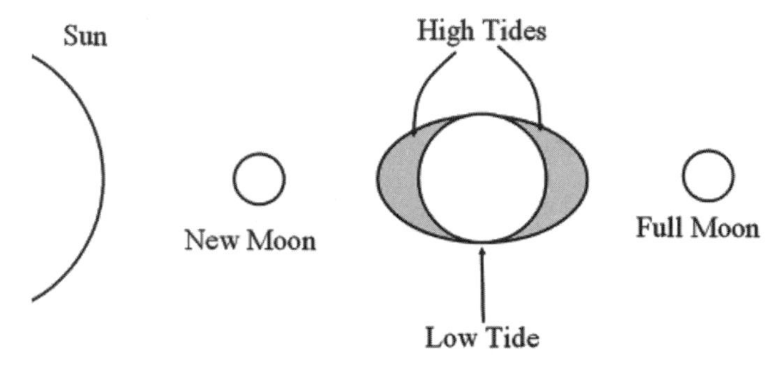

Figure 10.12. Spring tides

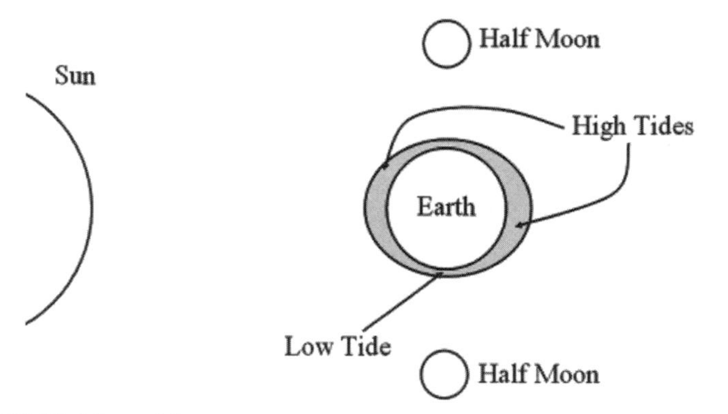

Figure 10.13. Neap tides

Lesson 10–4 Review

1. What is the name of the tide where the high tide does not get that high and the low tide does not get too low?

   a) slack b) ebb c) spring d) neap

2. Of the following lunar phases, which one causes the highest tides?

   a) waxing b) first quarter c) full moon d) waning

3. The balance point between the Earth and Moon is called the
   a) focus. b) epicenter. c) barycenter. d) penumbra.
4. The Moon has a bigger influence on our oceans than the Sun because
   a) the Sun is only made of gas and the Moon is solid.
   b) the Moon is closer to the Earth than the Sun.
   c) the water on the Moon attracts the water on the Earth.
   d) the magnetic attraction of the Moon makes the gravity stronger.
5. When the Sun, Earth, and Moon make a 180° angle, what kind of tide will we have?
   a) slack b) ebb c) spring d) neap

Lesson 10–5: Solar System

The Planets

The definition of a planet is currently under debate and has recently been looked at very closely. However, a good working definition of a **planet** is a large body that orbits the Sun. The word *large* is the main sticking point in the definition. Pluto is presently the topic of debate (Is it a planet or an asteroid?). Some circles include Pluto in the listing of planets while others have removed it.

The following chart covers the main facts about the planets in the solar system. There is a line separating Mars and Jupiter. This is because there is a big difference between the inner planets and the outer planets.

The inner planets are small rocky worlds similar to Earth. They have a high density becasue they are made of rock. Because the four inner planets are so similar to Earth, the inner planets are called Earth-like or **terrestrial planets**.

Saturn has a density lower than water—it will float in a large enough tub!

Starting at Jupiter, the planets are enormous (except Pluto, which will not be included in this group). The planets are many times larger than the Earth. They are primarily made of gases and have low densities. These planets are similar in size and density to Jupiter, so they are called **Jovian** (Jupiter-like) **planets** or **gas giants**.

Planetary Data Chart

	Distance from Sun *million km*	**Revolution Period**	**Eccentricity**	**Diameter** *km*	**Mass** *x Earth*	**Density**
Mercury	57.9	88 days	.206	4,880	.55	5.4
Venus	108.2	224 days	.007	12,104	.82	5.2
Earth	149.6	365¼ days	.017	12,756	1.0	5.5
Mars	227.9	687 days	.093	6,787	.11	3.9
Jupiter	778.3	11.86 years	.048	142,800	317.9	1.3
Saturn	1,427	29.46 years	.056	120,000	95.2	.7
Uranus	2,869	84 years	.047	51,800	14.5	1.2
Neptune	4,496	164.8 years	.009	49,500	17.2	1.7
Pluto	5,900	247.7 years	.250	2,300	.0025	2.0

Comets

Comets are chunks of ice and dirt that travel through space. They are remnants of the formation of the solar system. Early in the history of the solar system, comets were numerous an often crashed into the young planets. Most comets have now been swept out of the inner solar system by planetary collisions or deflections to the furthest reaches of our neighborhood. Although there is the occasional spectacle of a passing comet near to Earth, most comets now lie in a diffuse cloud at the furthest reaches of the Sun's influence in a spherical region called the **Oort Cloud**.

Thousands of astronomical units (AU) away from the Sun, the comets within the cloud are just barely held by the Sun's gravity. The slightest disturbance, from a collision or passing object, is enough to change a sleeping comet's path. In some cases, the comets are hurled towards the inner solar system.

For most of their lives, comets are dim "dirty snowballs," but as the comets get closer to the Sun, the frozen gases making up the body of the comet start to evaporate forming a cloud of gas called a **coma**. As the coma grows, it gets swept away from the sun by a stream of particles called the **solar wind** to form the characteristic *tail* of the comet. The tail of a comet acts much in the same way as a long coat would on a windy day. If the wind is in your face as you walk, your coat tails will billow behind you. However, if the wind is at your back, your coat will be blow in front of you. When the comet is moving away from the Sun, the tail actually is in front of the comet!

Most comets that we observe from Earth have highly elongated elliptical orbits. As they approach the Sun, they speed up and whip around the Sun only to be gone in weeks, never to be seen again for hundreds or even thousands of years. Haley's Comet, probably the most famous comet, has a fairly short period of 76 years, most of which is spent among the outer planets.

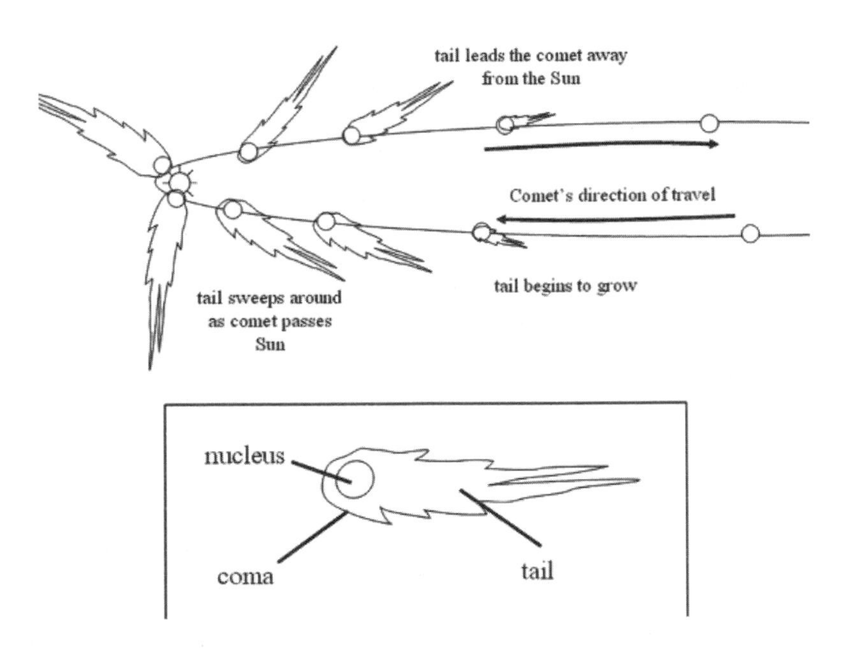

Figure 10.14. Comets

Meteors

Meteors are particles of dust and small rocks flying through space and have slightly different names, depending on exactly how they behave. While in space, the objects are called **meteoroids**. When a meteoroid enters the Earth's atmosphere, with speeds of several miles per second, the air friction heats the tiny rock until it vaporizes. The vaporization produces a streak of light across the sky that is called a **meteor,** or a "shooting star." If the meteor starts out large enough to survive the fiery passage through the atmosphere and slams into the ground, it then becomes a **meteorite**.

Meteoroids can randomly wander through space, but they can also be concentrated in bands of debris left behind by the melting of a comet. If the Earth should cross one of these bands of dust, we will experience a **meteor shower**. Meteor showers are actually fairly common events happening about once a month.

Name of Shower	Date of Peak Activity	Normal Duration	Expected Hourly Rates
Quadrantids	Jan 3	Jan 1–4	50
Alpha Aurigids	Feb 8	Feb 5–10	5
Virginids	Mar 20	Mar 5–Apr 2	5
Lyrids	Apr 21	Apr 19–23	10–15
Eta Aquarids	May 4	May 1–8	20
Ophiuchids	June 20	June 17–25	20
Delta Aquarids	July 29	July 26–31	25
Perseids	Aug 12	Aug 10–14	60
Kappa Cygnids	Aug 20	Aug 18–22	5–10
Alpha Aurigids	Sept 22	Sept 20–24	5
Giacobini-Zinner	Oct 9	Oct 7–11	Variable
Orionids	Oct 21	Oct 18–23	25
Taurids	Nov 1	Oct 15–Dec 1	15
Leonids	Nov 17	Nov 14–20	Variable
Geminids	Dec 13	Dec 10–15	50
Ursids	Dec 22	Dec 17–24	15

Meteor showers appear as if the meteors stream out from a central point in the sky. This is actually a trick of perspective. Just as the parallel tracks of a train appear to get closer the further they are from you, the meteor trails are actually parallel to each other. Meteor showers get their names from the constellation from which they appear to come.

Lesson 10–5 Review

1. Referring to the Planetary Data Chart, which planet's orbit has the biggest eccentricity?
   a) Mercury b) Earth c) Jupiter d) Pluto
2. Besides Pluto, which planet has the smallest diameter?
   a) Mercury b) Mars c) Earth d) Neptune
3. There used to be a planet at about 500 million kilometersfrom the sun, but it was destroyed. This doomed planet would have been between which two planets?
   a) Earth and Mars c) Saturn and Uranus
   b) Mars and Jupiter d) Neptune and Pluto
4. Terrestrial planets are
   a) low in density and made of gas.
   b) very large and made of gas.
   c) small and rocky.
   d) small and made of gas.
5. Jovian planets are all similar to
   a) Venus. b) Earth. c) Mars. d) Jupiter.
6. The tail of a comet comes from
   a) fiery exhaust from burning gases.
   b) dust and gas that are swept back by solar wind.
   c) small moons of the comet that follow it around space.
   d) long ice crystals formed by the comet moving quickly through the coldness of space.
7. An AU is
   a) a shorthand term for "the speed of light."
   b) the darker shadow during a solar eclipse.
   c) the average distance between the Earth and the Sun.
   d) a type of comet that comes from the Oort Cloud.

Lesson 10–6: Stars

Distances in Space

To measure the length of your desk, you might use a ruler marked with centimeters and come up with 150 cm or so from end to end. But to measure the length of a ball field, the centimeter becomes too small and difficult to count for such a long distance. The same problem holds true in space. The mile or kilometer (as large as they are) becomes too small to measure the distances involved in space. The next closest object to Earth, the Moon, is 220,000 miles away. To the next object after that, it becomes millions. The **AU**, discussed earlier, is the next yardstick used to measure distances. The size of the average distance between the Earth and Sun, the AU makes a good unit of length to measure distances within the solar system.

Once out of the solar system, even the AU becomes too small so a larger unit of length is needed: the light-year. A **light-year** is the distance that light travels in one year. At 300,000 km/second (186,000 miles/sec) light will travel about 9.5 trillion kilometers (5.9 trillion miles) in a year. This sounds like a very big distance but it still can become small in space. The next closest star to our Sun is Proxima Centauri, which is 4.6 light-years away. You will have to travel at the speed of light for 4.6 years to reach it.

A light-year is NOT a measurement of time.

Bad science fiction will have people discuss their futuristic, spacey ages in terms of how many light-years old they are.

There is still one more unit of distance that is bigger than a light-year: the **parsec**. The parsec uses a concept called **parallax**. A simple way of explaining parallax is to point your finger at a distant object. Close one eye, then switch eyes and you will see your finger appear to jump left and right. The jumping of your finger is caused by the distance between your eyes giving each one a slightly different angle between your eye, finger, and background objects. As Earth revolves around the Sun, we see near stars at different angles against the background of distant stars (see figure 10.15).

The **parsec** is the distance at which an object will be if it appears to shift by 1 second of arc (1") because of Earth's trip around the Sun. This equals a distance of about 3.26 light-years.

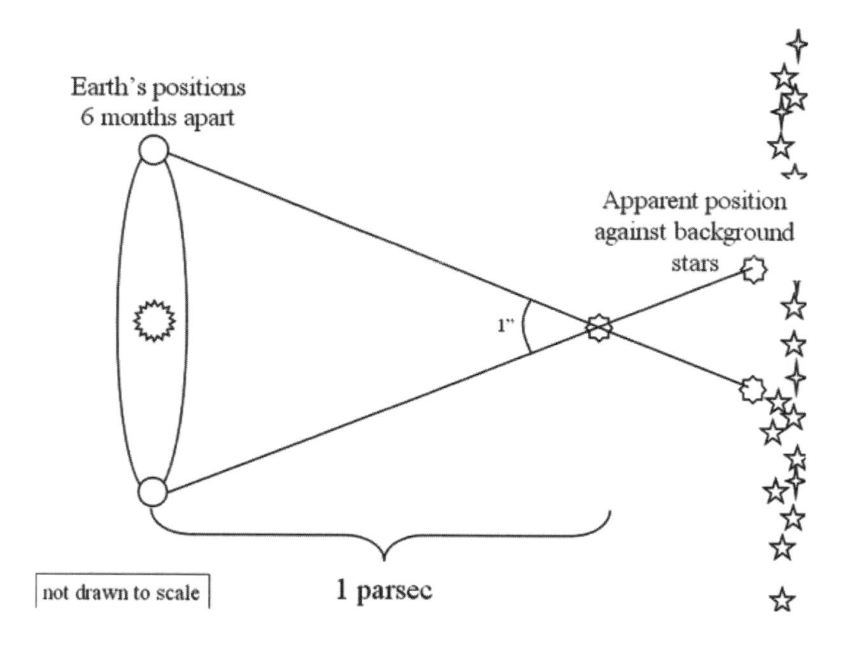

Figure 10.15 Parallax

Constellations

When we look into the night sky, the stars appear to be randomly scattered throughout the blackness. A natural tendency is to "connect the dots" and draw pictures. The pictures that ancient people drew are still used today to refer to groupings of stars. Very often it is difficult to "see" the image, such as a crab, in the outline of the stars. This is simply because not everyone has the same imagination. Plus, there is much more light at night now, which makes it harder to see all of the stars. The constellations are used to describe not only patterns of stars, but also regions or territories in our sky.

Constellations are groups of stars as we see them from Earth. They are not grouped together in space. In fact, the groups of stars in a constellation may be very far apart from each other, but they line up from our point of view. It's like watching two planes fly past each other: They don't crash because one is further away than the other, but they seem to just miss each other.

The most famous star in the sky is **Polaris**, otherwise known as the **North Star**. Its fame doesn't come from its brightness, color, or size. In fact, there's nothing special about Polaris except the fact that it happens to be lined up with Earth's axis of rotation. Other than that, it is just a regular old star.

Because Polaris lines up with the axis of rotation, it never appears to move as the rest of the sky appears to spin as the Earth rotates. Think of Polaris as a pin stuck into the sky and of the rest of the heavens slowly spinning around the pin. One way of seeing the rotation of the sky around Polaris is to take a special long-time photograph. Due to Earth's rotation, a long-exposure photo will cause the stars to slowly streak across the film resulting in **star trails**.

When looking north, the star trails will make a bulls-eye pattern with Polaris at the center. When looking to the east, the stars will trail upward and slant towards the south—following the same path as the Sun and moon. To the south, the trails make sweeping arcs. And to the west, the stars set slanting from the south down into the horizon.

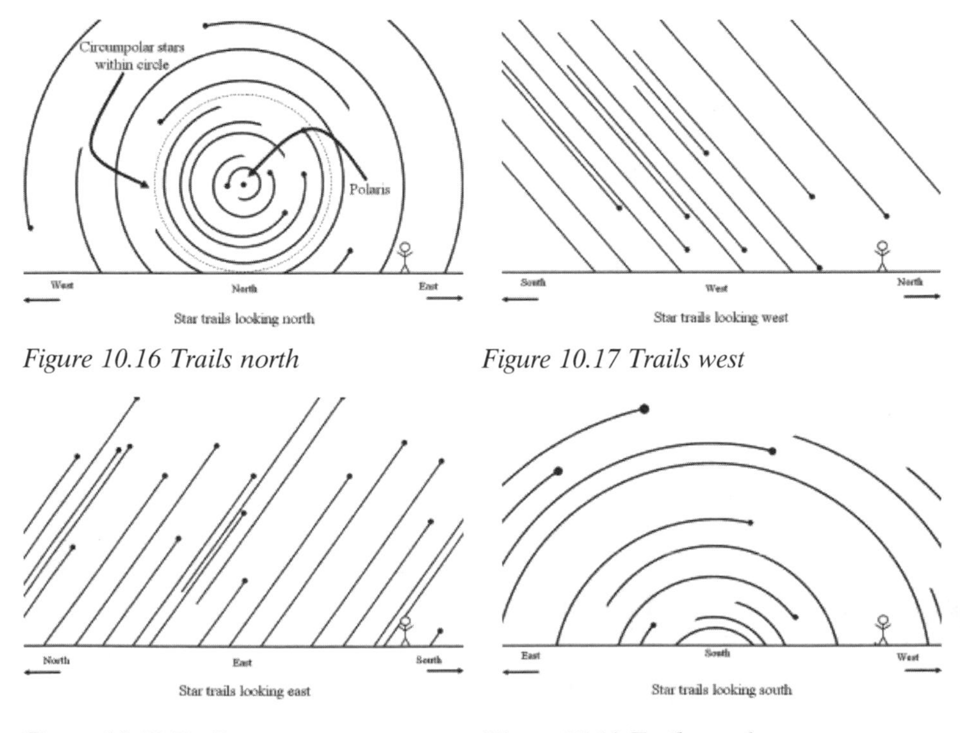

Figure 10.16 Trails north

Figure 10.17 Trails west

Figure 10.18 Trails east

Figure 10.19 Trails south

Circumpolar Stars

If you look closely at the diagrams of the star trails looking north, you will see that there are some stars that have such a tight circle that they never dip below the horizon (stars within the dotted circle). These stars will be visible all night. In fact, these stars will be visible every night of the year. The stars that never set because they are so close to Polaris are called **circumpolar stars**. The constellations that they are in are **circumpolar constellations**. The circumpolar stars vary depending on your latitude, but, in most of the United States, bright constellations such as the Big Dipper (Ursa Major), the Little Dipper (Ursa Minor), and Cassiopeia are visible every night of the year.

The Workings of a Star

A **star** is a very large ball of gas, mainly hydrogen and helium that glows from its own energy. The Sun is a typical star of average size and temperature, just like thousands and millions of others in space. In the center of a star, the incredible heat and pressure actually squeezes hydrogen atoms together to form helium in a process called fusion. Four hydrogen atoms (with an atomic mass of 1.008 each) combine to form one helium atom (with an atomic mass of 4.026). But, when the helium is formed, there is a tiny bit of leftover mass:

four hydrogen atoms:	$1.008 \times 4 = 4.032$
one helium atom:	4.0026
leftover mass:	$4.032 - 4.0026 = .0294$

$E=mc^2$

E= energy

m= mass

c^2 = the speed of light squared (300,000 km/sec × 300,000 km/sec)

According to Einstein's famous equation($E=mc^2$) the leftover mass is converted into energy—a lot of it. This creates a nuclear explosion. The notion that a star is a ball of fire is technically inaccurate. It is not burning fuel, which is a chemical reaction; it is fusing atoms, which is a nuclear reaction. And the Sun is converting tons of hydrogen into helium every second! Essentially, the Sun is a nuclear reaction that has continued for the last four-and-a-half billion years.

The Life of a Star

A star begins as a cloud of dust and gas floating in space. The gravitational attraction of each speck of dust and each atom of gas starts to pull the cloud in on itself. The contraction of the cloud continues and gets faster and stronger as the cloud gets more compacted. Friction between the particles creates heat and the temperature increases. When the temperature gets hot enough and the density of the cloud gets high enough, nuclear fusion will begin and the star will ignite: it is born.

From this point on, the star will convert hydrogen into helium for the majority of its life. The bigger and more massive the star, the stronger the gravity it generates. With more gravity comes more pressure and faster nuclear reactions—the star "burns" hotter and uses its fuel faster. The star will reach an equilibrium between gravity, which tries to collapse the star inwards, and the expansion forces from the nuclear explosions. During this phase of the star's life, it is stable and remains this way for billions of years. It is called a "**main sequence star**," which means that it is essentially an adult star.

Once the star consumes its primary source of fuel, hydrogen, it will begin to use up the next heaviest element on the periodic chart—helium. At this point, the star starts reaching the end of its life and becomes unstable. The collapsing force of gravity and explosive forces become unbalanced and the star will swell many times its size into a red giant. A **red giant** is a very large but cool star. When our Sun swells up into a red giant (in about 5 billion years) it will become large enough to consume Earth and maybe even Mars!

At this point, the destiny of the star depends on the mass with which it started. If the star started with a size of 7 times that of our Sun or smaller, it will stay as a red giant until the fuel is used up, at which point the star will rapidly collapse while throwing off its outer layers and shrink into a white dwarf that will eventually cool and die. A **white dwarf** is a very hot but small star that could be the size of Earth or smaller.

If the star started out larger than 7 times the size of our Sun, it will have a more dramatic finish. After the red giant stage, the collapse of the star will be catastrophic because of the intense gravity that massive stars have. When it collapses, the star will reach a point where the matter gets so compressed that the outer layers bounce off of the inner and the star explodes. The brilliant explosion is called a **supernova**. After the supernova, the outer layers of the star will dissipate into space as a cloud of gas and dust (to begin the cycle again). The very small but extremely dense core of the star will be left as a pulsar. A **pulsar** is a tiny star (smaller than the Earth) that spins many times a second. The light from a spinning pulsar appears to flash on and off (or pulse) several times a second, which is where it gets its name.

If the original star is massive enough, the collapse from the red giant stage gets out of control. As the star collapses, the atoms get closer, which makes the gravity stronger, which makes the atoms gets closer. This vicious circle doesn't stop and the entire mass of the star is compressed into a single point—a singularity. Literally, the entire mass of a star that was originally more than 7 times the size of the Sun would fit in the palm of your hand! Gravity gets so magnified that nothing can pull away from the star—not even light. It becomes a black hole.

Classification of Stars

Stars are classified by their *color* and *luminosity*. The **luminosity** of a star is a measure of how much light is given off by the star. One of the most significant factors for luminosity is the size of the star. If all other things were equal, a larger star would have a larger surface area emitting light. The Sun is used as the standard and has a value of 1. A star that is 10 times brighter than the Sun has a luminosity of 10 and so on.

The temperature of a star is also important in classification. Even though we cannot fly to any star, including our own, to stick a thermometer into it, we can measure the temperature by the color of its light. Imagine a red-hot piece of iron. It gives off light only because of its heat, not because of some chemical reaction such as fire. If it gets hotter, it changes from red, to orange, to yellow and may even become white-hot. A star's temperature is measured in much the same way. Red stars are the coolest. Yellow, as the Sun, is medium. White is very hot, and blue-white is the hottest.

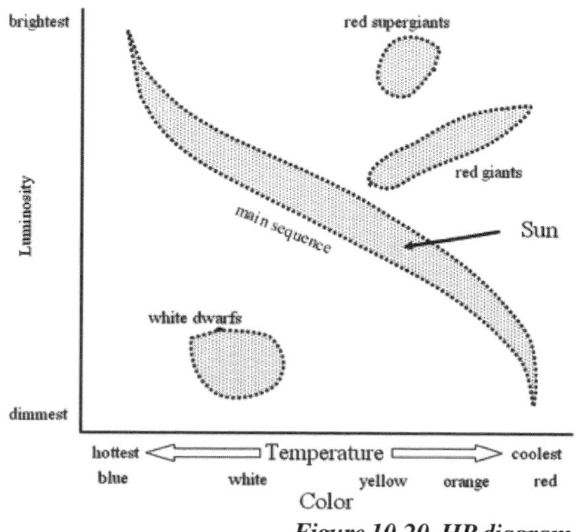

Figure 10.20. HR diagram

The Hertzsprung-Russell Diagram (often referred to as the "HR Diagram") is a graph of luminosity and temperature/color of all visible stars. When the graph is filled with the stars, they fall into distinct patterns: the main sequence, white dwarfs, red giants, and red supergiants.

Lesson 10–6 Review

1. A constellation is
   a) a group of stars that make a picture in the sky.
   b) a set of planets.
   c) all of the brightest stars in the sky.
   d) any star that does not set because it is so close to Polaris.

2. A light-year is
   a) the amount of time it takes a star to revolve once.
   b) the metric version of an AU.
   c) the distance light will travel in a year.
   d) the speed of light in scientific notation.

3. The majority of a star's life is spent in which stage?
   a) main sequence
   b) red giant
   c) white dwarf
   d) red supergiants

4. Which direction does this picture of star trail picture represent?

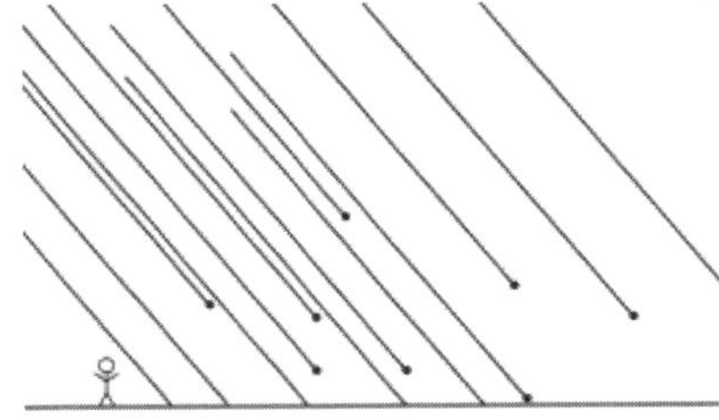

   a) north
   b) south
   c) east
   d) west

5. A good unit to measure the distances between planets would be
   a) kilometers.
   b) astronomical units.
   c) light-years.
   d) parsecs.

6. When a star is luminous, it is
   a) dense.
   b) nearby.
   c) in the process of dying.
   d) bright.

Lesson 10–7: Beyond the Stars

Spectral Lines

We know that distant stars are made of hydrogen and helium. How do we now this? Here's how. When white light is viewed through a prism it makes a rainbow called a **spectrum**. When an element is heated, it begins to glow and gives off light. When an element's light is viewed through a prism, you don't see a full spectrum; you only see a few stripes of color. This pattern of stripes is unique for that element, and no other element will make the same pattern. Therefore, the spectral lines for an element can be used as a fingerprint to identify the element.

Helium was discovered on the Sun before it was found on Earth.

When the Sun's spectrum is viewed, there are many lines from the many elements in the Sun. When each line is matched with the element that created it, the various elements can be identified. When this technique was first used, there was a set of lines that matched no known element. A new element was discovered. As a result the new element was named after the Helios, which is Greek for Sun. It is the "sun element" heli-um.

Doppler Shift

When a train speeds past you, blowing its whistle, you will hear a shift in the pitch of the whistle. As the train approaches you, the sound waves in front of it get compressed, which makes a slightly higher pitch. After it passes, the sound waves get stretched out and you hear a lower pitch. All together it will make a "wheee-oooo" sound. The effect gets more exaggerated with greater speed. First studied by a scientist named Doppler, this shifting sound is called the **Doppler Effect** or "Doppler shift."

The spectral lines from a star go through a similar shifting. If the star is coming closer, the fingerprint lines will stay together but get shifted toward the blue end of the spectrum. If the star is moving away, the lines will shift toward the red end. Just like the shifting in sound, the shift of the light gets more exaggerated with greater speed.

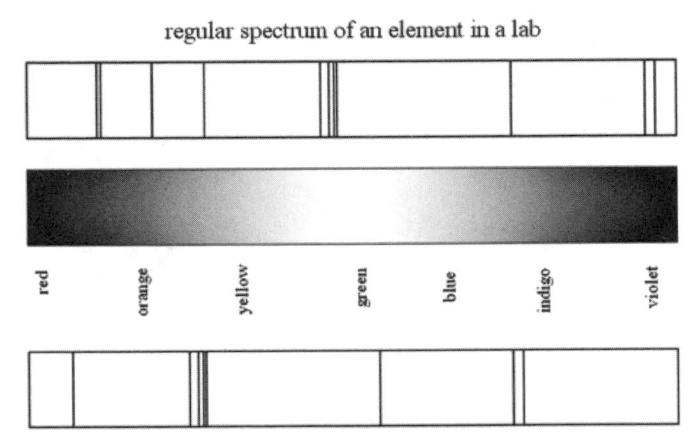

The Doppler Effect led to two significant discoveries. The first was that all distant galaxies are moving away from us; they all have a red shift. The only way that this can be explained is that the Universe is expanding. The second discovery was that, the further away an object is from us, the faster it is moving.

If these two pieces of information are put together, it implies that not only is the Universe expanding, but it is exploding! If the "video tape" of the Universe's movement is reversed, all of the observable galaxies would appear to get closer and converge at a single point. The convergence would have happened about 15 billion years ago, which is thought to have been the moment of creation of the Universe.

The Big Bang

It is widely accepted that the moment of creation was an event called The **Big Bang**. At this moment, for a still unknown reason, a point in empty space exploded. In that explosion, everything was created—everything! Every proton and electron, every shred of energy was created at that moment. From that moment on, the Universe exploded outward and the explosion continues today. This explains the expanding Universe. It also explains why the farthest galaxies are moving faster.

The big debate today is the eventual fate of the Universe. Will it continue to expand forever? Will it reach a certain size and then stop expanding? Will it stop expanding and then begin to fall back into itself and get smaller until it all crunches back together? The simple answer to these questions is: It depends on how much matter there is in the Universe.

If there is "too much" matter, then the gravity from all the mass will cause the Universe to contract. If there is "too little" matter, then there will not be enough gravity to slow the expansion, and the Universe will expand forever. Right now, there is not enough information to tell which way it will go.

The Order of the Universe

Going from smallest to largest, the Universe is arranged at several levels. The smallest "unit" is the **star**. Stars can be found by themselves, they can be paired, or there can even be multiple stars orbiting around each other as small systems.

It appears that there is a good fraction of stars that have planets making the next largest grouping: the **solar system**. Recent discoveries confirm the existence of many stars with planets orbiting around them. It seems as if stars with their own solar systems are fairly common.

A galaxy is the next level or order. A **galaxy** is a collection of billions to trillions of stars. It is held together by the mutual gravity of all the stars within it. Our galaxy, for instance, is the Milky Way galaxy. It is 100,000 light-years across and contains trillions of stars—one of which is our Sun. Galaxies come in many shapes and sizes. Like our Milky Way, they can be spiral-shaped, or they can be elliptical or irregular. Deep space photographs have shown that there are just as many (or even more!) galaxies in space as there are stars in the Milky Way.

The last level that we are aware of is the **Universe**. The Universe literally means "everything." Everything there is, is contained in the Universe. So far, the estimate is that there are more galaxies in the Universe than there are stars in a galaxy. An infinite number of stars is not a bad estimation. Oh, the possibilities!

Life in the Universe?

One big question often asked at this point is: "Is there life out there?" We haven't confirmed it yet, but there are some promising leads. One of the first steps to scientifically pursuing this question was the development of the "**Drake Equation**." Essentially, the Drake Equation estimates the odds of life in the Universe based on the odds of stars that have planets, how many planets are at a habitable distance from their star, how old the planets are, and other factors that are necessary for life to develop. Unfortunately, most of the numbers are merely estimates because Earth is the only model upon which to base the numbers.

Despite the guesswork involved in using the Drake Equation, there are a few things worth considering. We have discovered life right here on Earth in places that were believed to be too inhospitable an environment to sustain life. Life has been found in extreme cold, heat, and dry climates. Some of the most extreme life that was recently discovered lives in super hot (above boiling temperatures), mineral-rich water spewing from underwater volcanic vents. The life thrives on the chemical nutrients of the vents. This discovery redefined what we consider to be a suitable environment because the organisms are based not on photosynthesis, but *chemosynthesis*—the chemicals as nutrients. Considering how tenacious life has been on our planet, our horizons of exploration now consider far more possibilities than ever. With this in mind, and considering that there are essentially an infinite number of stars in the Universe, many of which have planets, the probability of life is promising.

Lesson 10–7 Review

1. Draw these symbols inside each other with the smaller objects inside the larger objects. (Hint: Do not group the shapes in the sizes that they are drawn.)

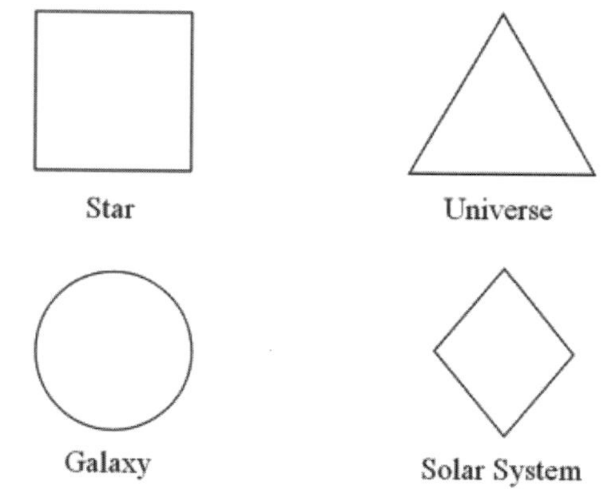

2. The Big Bang is
   a) the death explosion of a star.
   b) a nickname for Einstein's equation.
   c) a collision between two stars.
   d) the theory of how the Universe was created.

3. The changing frequency of a train whistle is called the
   a) Coriolis Effect.
   b) Greenhouse Effect.
   c) Doppler Effect.
   d) Locomotive Effect.
4. A red shift in a star's spectrum means that the star is
   a) moving towards us.
   b) moving fast, but we can't tell what direction.
   c) moving away from us.
   d) moving slow, but we can't tell what direction.
5. How many galaxies are there in the Universe?
   a) dozens
   b) hundreds
   c) thousands
   d) trillions
6. Which of the following is evidence that the Universe is expanding?
   a) We can see stars exploding every once in a while.
   b) All galaxies had a red Doppler shift.
   c) Stars rise in the east and set in the west.
   d) Stars convert matter into energy using $E=mc^2$.

Chapter Exam

1. Which of the following ellipses is least eccentric?
   a)
   c) 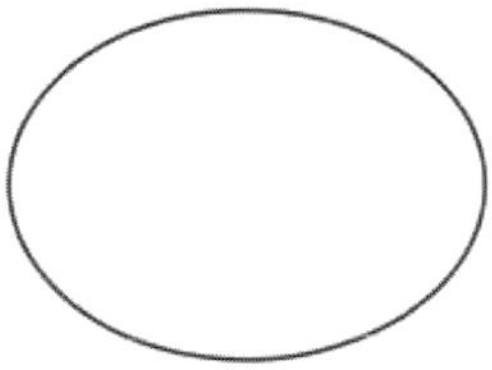
   b)
   d)

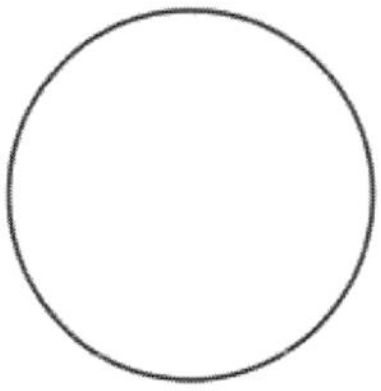

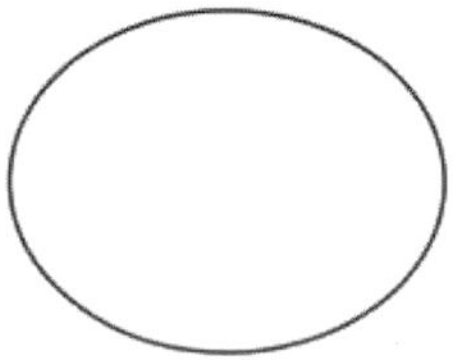

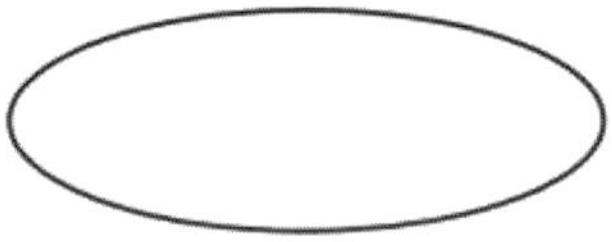

2. A bigger red shift means
   a) the star is larger.
   b) the star is a cool temperature.
   c) the star is moving away faster.
   d) the star is near the end of its life.

3. An ellipse with an eccentricity of 1.0 will have what shape?
   a) a perfect circle
   B a straight line
   C a slightly squashed ellipse exactly like the Earth's orbit
   d) an oblate sphereoid

4. Refer to the Planetary Data Chart (on page 308) for the following question. Which of the following planets has an eccentricity less than that of the Earth's orbit?
   a) Mercury b) Jupiter c) Pluto d) Venus

5. Which graph shows the relationship between luminosity and size?

a)

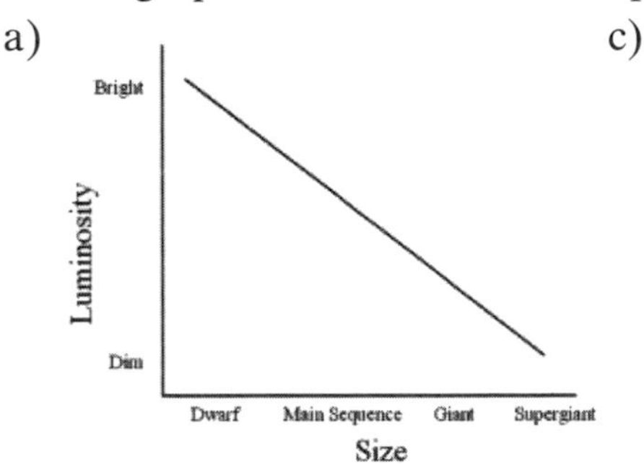

c)

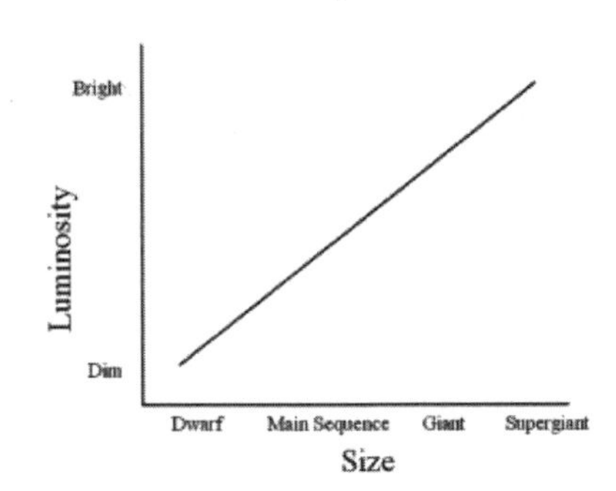

b)

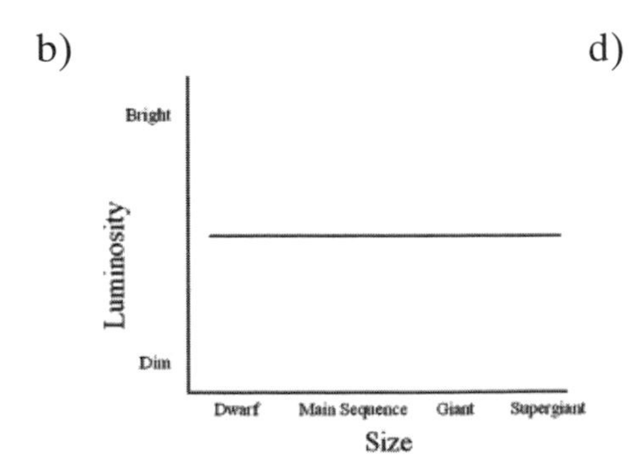

d)

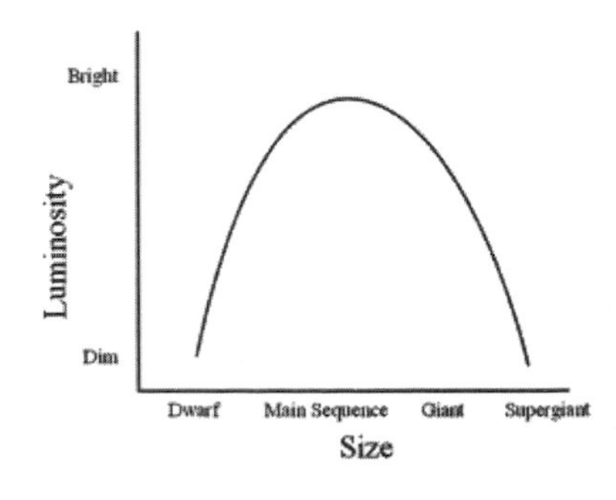

6. What effect does decreasing the distance between the foci have on the shape of the ellipse?
   a) Eccentricity increase.
   b) Eccentricity decreases.
   c) Eccentricity stays the same.
   d) Eccentricity does not change.

7. How does the velocity of this comet change as it moves around the Sun from position A to position D?

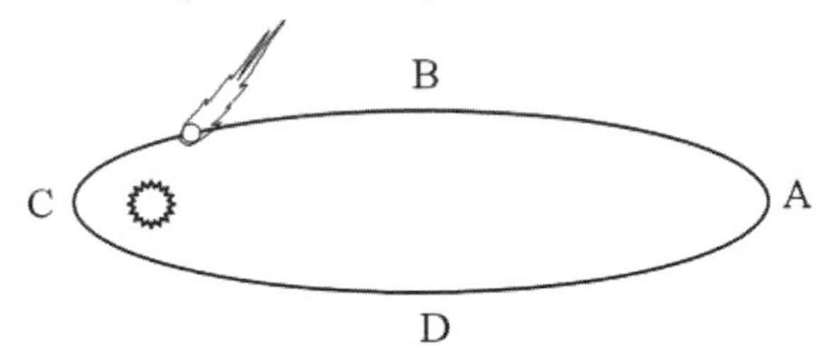

   a) Its velocity constantly increases from A to D.
   b) Its velocity constantly decreases from A to D.
   c) Its velocity decreases from A to C and then increases.
   d) Its velocity increases from A to C and then decreases.

8. Tides are caused by
   a) the alignment of the Earth, Sun, and Moon.
   b) our distance from the Sun.
   c) our speed around the Sun.
   d) a tsunami.

9. Which event is the most rare?
   a) a lunar eclipse
   b) a solar eclipse
   c) a spring tide
   d) a new moon

10. The two largest planets in the solar system are
    a) Earth and Mars.
    b) Uranus and Neptune.
    c) Mercury and Pluto.
    d) Saturn and Jupiter.

11. The nearest star to our solar system is only a few
    a) million kilometers away.
    b) AUs away.
    c) light-years away.
    d) micrometers away.

12. A stars begins as
    a) a planet.
    b) a cloud of gas.
    c) an atom.
    d) a group of comets.

Answer Key

Answers Explained Lesson 10–1

1. **D** is **correct** because it is the most squashed.
2. **A** is **correct** because the Earth's orbit is so close to a perfect circle that it cannot be told apart from one. (Ellipse A could also have an eccentricity of .017 and you just can't see the distortion!)
3. **C** is **correct** because 4 cm ÷ 10 cm = .4
4. **A** is **correct** because spreading out the foci will stretch out the ellipse.
5. **D** is **correct** because the foci will have the same distance apart as the width of the ellipse. For example, 10 cm ÷ 10 cm = 1.
6. **B correct**. The Sun is off to one side of the Earth's orbit, which explains why we get closer to—and further fro—the Sun throughout the year.
7. **C** is **correct**. A perfect circle has both foci at on point. The distance between the foci is zero. Zero divided by any number gives an eccentricity of zero.
8. **A** is correct. The Earth's orbit looks like **a circle**. The orbit is so close that it appears to us as if it were a circle.

Answers Explained Lesson 10–2

1. **B** is **correct** because waxing means to shine like a waxed floor.

   D is incorrect because waning means to get dimmer.

   A and **C** are incorrect.

2. **A** is incorrect because the Moon can be seen at some time during the day. You've probably seen it in the afternoon or near sunset.

   B is incorrect because sometimes the Moon is on the night-time side of the Earth.

   C is **correct**. It just isn't as obvious during the day against the blue sky and white clouds.

   D is incorrect.

3. **C** is **correct** because the Moon rises 50 minutes later each night.
4. **A** is incorrect because the Moon can be seen during the day.

   B is incorrect because at a New Moon you cannot see the Moon even if you looked directly at it.

   C is incorrect because an eclipse of the Moon happens during the full moon phase. (It is not a solar eclipse either.)

   D is **correct**.

5. **A** is incorrect because asteroids are small bodies that orbit the Sun.

   B is incorrect because planets are large bodies that orbit the Sun.

   C is **correct**.

   D is incorrect because a meteor is a very small rock.

6. **A** is **correct**.

Answers Explained Lesson 10–3

1. **A** is incorrect because *solar* eclipses happen during the New Moon.

   B is **correct** because the Moon needs the Earth to be in between itself and the Sun for a lunar eclipse.

   C and **D** are incorrect.

2. **A** is **correct**. The tilt takes the Moon out of direct alignment most of the time.

   B is incorrect. It is true that the Earth's axis of rotation is tilted, but it doesn't affect eclipses.

   C is incorrect.

   D is incorrect because we would still have eclipses. The Moon would just not be large enough to cover the Sun's disc.

3. **A** is incorrect because you need to be in the umbra to see a total eclipse.

   B is **correct**.

   C is incorrect because some of the Sun will be blocked to cause the partial shadow.

   D is incorrect.

4. **A** is incorrect because a partial eclipse happens when there is a "bite taken out of the Sun." In this case, there will be a ring around the dark shape of the moon.

   B is incorrect because the Sun will not be 100% covered.

   C is **correct**. An annular eclipse is caused by the Moon being too small and far away to completely cover the Sun's disc.

   D is incorrect.

5. **A** is **correct**. The Earth's shadow is much larger than the entire Moon, so there is a better chance that the Moon will cross into it.

Answers Explained Lesson 10–4

1. **D** is **correct**.

2. **A** is incorrect because waxing is any phase that happens while the Moon is getting brighter from night to night.

   B is incorrect because during the first quarter, the Earth, Sun, and Moon are at 90° angles and the gravitation pull on the oceans is weakest.

C is **correct** because during a full and new moon the Earth, Sun, and Moon are lined up.

D is incorrect because waning is any phase that happens while the Moon is getting dimmer from night to night.

3. **A** is incorrect because the focus is one of the two centers of an ellipse.

   B is incorrect because an epicenter is an earthquake term.

   C is **correct**.

   D is incorrect because the penumbra is a partial shadow.

4. **A** is incorrect. Even though the Sun is made of gas, it is much more massive, which is more important for gravitational pull.

   B is **correct** because distance between objects is important for gravitational pull.

   C and **D** are incorrect.

5. **A** and **B** are incorrect.

   C is **correct**. The straight line allows the gravity of the Sun and Moon to add up.

   D is incorrect because a neap tide happens when the Earth, Sun, and Moon make a 90° angle.

Answers Explained Lesson 10–5

1. **D** is **correct** because Pluto has an eccentricity of .25.
2. **A** is **correct** because Mercury has a diameter of 4,880 km.
3. **B** is **correct** because the distance of 500 million kilometers is between the distances of Mars and Jupiter.
4. **A** is incorrect because the gas giants are low in density and made of gas.

   C is **correct**.

   B and **D** are incorrect.

5. **A**, **B,** and **C** are incorrect because they are terrestrial or Earth-like planets.

   D is **correct** because Jovian means "Jupiter-like."

6. **B** is **correct**.
7. **C** is **correct**. AU stands for "astronomical unit," which is the average distance between the Earth and Sun.

   D is **incorrect**.

Answers Explained Lesson 10–6

1. **A** is **correct**.

   B is incorrect because a set of planets is a solar system.

C is incorrect.

D is incorrect because any star that is that close to Polaris is a circumpolar star or are part of circumpolar constellations.

2. **A** is incorrect because that is an orbital period or a year.

   B is incorrect.

   C is **correct**.

   D is incorrect.

3. **A** is **correct** because the main sequence is the long lasting, stable portion of a star's life.

   B, **C,** and **D** are incorrect.

4. **D** is **correct**. The stars are coming from the upper left down to the lower right. To help understand, use your finger to point to the eastern horizon where the stars rise. Make a big arc towards the south while turning to the west. You will see that when you turn to the west. Your finger will trace the same lines as in the picture.

5. **A** is incorrect because kilometers are too small.

   B is **correct** because the largest distance in the solar system is only 40 AUs—a good number to work with.

   C is incorrect because the solar system is not even a fraction of a light-year across. Pluto is a couple of light-hours away!

   D is incorrect because a parsec is too large.

6. **D** is **correct**. Luminosity means the amount of light a star emits.

Answers Explained Lesson 10–7

1.

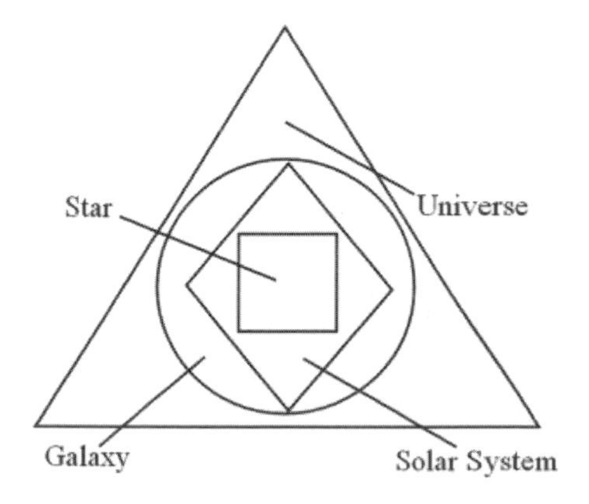

2. **A** is incorrect because a supernova is the death explosion of a star.

   B and **C** are incorrect.

   D is **correct**.

3. **A** is incorrect because the Coriolis effect is the turning to the right caused by the Earth's rotation

   B is incorrect because greenhouse effect is the trapping of heat by carbon dioxide.

   C is **correct**.

   D is incorrect.

4. **C** is **correct**. As the star moves away, the light waves get stretched, which shifts the colors towards the red.

5. **D** is **correct**. With more powerful telescopes, more and more galaxies are observed—too many to count.

6. **A** is incorrect because stars explode as a normal part of their life cycle.

   B is **correct** because a red shift means the galaxy is moving away. The only way for all galaxies to have a red shift is if the Universe is expanding in all directions.

   C is incorrect because the rising and setting is caused by Earth's rotation.

   D is incorrect. It doesn't answer the question.

Answers Explained Chapter Exam

1. **A** is **correct** because it is the least warped.

2. **A** is incorrect.

   B is incorrect because a red color is a cool star, but a red shift means it is moving away.

   C is **correct** because a red shift means the star is moving away. A bigger shift is a bigger speed.

   D is incorrect.

3. **A** is incorrect because a perfect circle has an eccentricity of zero.

   B is **correct**.

   C and **D** are incorrect.

4. **D** is **correct** because Venus has an eccentricity of .007 while the Earth's is .017.

5. **C** is **correct** because the bigger stars have a greater luminosity.

6. **B** is **correct** because the closer the foci are, the closer to a perfect circle the ellipse becomes.

7. **C** is **correct**. From A to C the comet gets closer to the Sun. Gravity is stronger so the comet travels faster. After C, the comet moves away and the Sun's gravity slows it down.

8. **A** is **correct** because the Sun and Moon pull on the oceans strongest when they are lined up.

   B is incorrect because our distance from the Sun does not change enough to affect the oceans.

   C is incorrect because our speed around the Sun does not affect the oceans.

   D is incorrect. A tsunami is called a tidal wave but has nothing to do with tides. It is caused by earthquakes.

9. **A** is incorrect because the Earth's shadow is so large that there's a good chance that the Moon will cross into it.

   B is **correct**.

   C is incorrect because a spring tide happens twice a month.

   D is incorrect because a new moon happens once a month.

10. **D** is **correct**.

11. **A** is incorrect.

    B is incorrect because AUs are too small to measure the distances to the stars.

    C is **correct**. Proxima Centauri is 4.6 light-years away.

    D is incorrect because micrometers are tint fractions of a meter.

12. **A** is incorrect because stars and their planets form at about the same time.

    B is **correct**. Stars form as clouds of gas contract.

    C and **D** are incorrect.

Glossary

Abrasion: when two rocks scratch each other by rubbing against each other.

Absolute time: measuring the time of an event using a number on the time line.

Absolute zero: the coldest that any matter can be. It is –273° Celsius.

Adiabatic cooling: air cooling as it rises due to expansion and spreading the heat out.

Adiabatic heating: the heating of air caused by pressurization, which concentrates the heat.

Adiabatic lapse rate: the rate at which dry air cools as you go up in altitude.

Agent of erosion: anything that pushes sediment from one location to another.

Air pressure: weight of the air above you.

Albedo: a ratio between the amount of sunlight that comes in as compared to the amount of sunlight that is reflected.

Alberta clipper: a strong but fast, moving low pressure system.

Alpine glacier: a "river of ice" that flows down through mountain valleys.

Angle of insolation: the angle at which sunlight strikes the ground.

Anticyclone: high pressure system with an outward and clockwise air flow.

Apparent motion: the way an object in space appears to move caused by the motion of the Earth.

Aquitard: a layer of low permeability that acts as a cap.

Artesian well: a source of water where the water comes out of the ground under natural pressure.

Asthenosphere: the upper mantle in which convection takes place.

Atmosphere: the layer of air that surrounds the Earth.

Basal cleavage: splitting into flat sheets.

Benchmark: a spot on the surface of the Earth where an exact elevation was measured.

Bioclastic: "fragments of life" rock made up of broken shells.

Caldera: the wide crater at the top of a shield volcano.

Capacity (humidity): the amount of water vapor that the air can hold.

Capillarity: water being drawn upwards against the pull of gravity through tiny pores.

Capillary fringe: a zone right on top of the water table where capillary action draws some water upwards.

Cementation: the process of "gluing" sediments together.

Chemical weathering: the breaking down of rock material by the reaction of some chemical.

Chinook: a warm dry air that descends on the leeward side of a mountain.

Cinder cone: a volcanic mountain made from cinders.

Cinder: small pieces of volcanic rock thrown from a volcano.

Clastic: a sedimentary rock made up of bits and pieces of any other rock.

Cleavage: the property of a mineral causing it to split along flat surfaces.

Climate: the typical type of weather that an area has.

Cloud base height: the altitude at which clouds will form. Air temperature and dew point temperature are equal.

Cold front: the leading edge of an incoming cold air mass.

Compaction: to compress under great pressure.

Composite cone: volcanic mountain made from alternating lava flows and cinder eruptions.

Conchoidal fracture: breaking into bold-shaped depressions.

Conduction: heat transfer between two materials that are touching.

Cone of depression: a section of the water table that is drawn down as the water is pumped out.

Contact metamorphism: a metamorphic change in rock caused by contact with extreme heat.

Continental air mass: a dry air mass.

Continental drift: the theory that the Earth's land masses are slowly moving around.

Continental glacier: a large ice sheet that covers entire continents.

Contour interval: the amount of elevation change from one contour line to the next.

Convection cell: a complete loop caused by rising and sinking air.

Convection current: the movement of air caused by the density differences in hot and cold air.

Convection: a heat transfer method that is caused by hot air rising and cold air sinking.

Converges: air currents come together caused by the sucking action of a low pressure system.

Coriolis Effect: wind and ocean currents are deflected to their right in the Northern Hemisphere caused by the Earth's rotation.

Correlation: matching one layer of rock with another layer of rock in a distant location.

Cross-beds: slanted layers of sorted sediments that are deposited in river deltas or meanders.

Crust: the outer rock layer of the Earth on which we live.

Crustal activity: any movement of the crust, such as earthquakes, mountain building, and volcanoes.

Cubic cleavage: breaking into cubes where each of the corners meet at 90 degrees.

Cyclic: a pattern that repeats over and over again.

Cyclone: a low pressure system that has an inward and counter-clockwise air flow.

Delta: the triangular-shaped deposit at the mouth of a stream.

Density: the amount of matter that is packed into each cube of space.

Dew point lapse rate: the rate at which the temperature of humid air cools as you go up in elevation.

Dew point: the temperature at which humidity becomes 100% and condensation forms.

Direct relationship: a relationship between two numbers where as one number increases, the other number also increases. Graphs of direct relationships will slant upwards to the right.

Discharge: the amount of water flowing through a stream.

Divergent plate boundary: the place where two plates are spreading apart.

Divergent: when currents of air flow away from each other caused by high pressure.

Downhill creep: the movement of soil sliding downhill slowly over many years.

Dynamic equilibrium: a system in which two opposite processes balance each other so that there is no overall change.

Eccentricity: a measure of how squashed an ellipse is.

Elastic rebound theory: theory that says that, before an earthquake, the rocks will stretch as far as they can and then suddenly snap.

Ellipse: a squashed circle. The locus of points that are equidistant from two foci.

Epicenter: the spot on the surface of the Earth closest to the source of the earthquake.

Erosion: the transportation of sediments from one location to another.

Erratic: a boulder transported by a glacier.

Evaporites: sedimentary rock formed by minerals deposited after the evaporation of water.

Evapotranspiration: the combination of evaporation and transpiration.

Evolution: the gradual changing pattern of organisms.

Exfoliation: the flaking of the outer surface of a rock.

Extinction: the final ending of a species.

Extrapolate: to take a set of data and extend it into the future.

Extrusive: igneous rock formed on the surface of the Earth.

Fault: a crack in the Earth where the movement of an earthquake takes place.

Field: a section of the Earth where a measurement can be taken at any point.

Flood plain: the wide flat section of land next to a river.

Foci: the plural of focus.

Focus (ellipse): one of the two central points of an ellipse.

Focus: the spot within the Earth where an earthquake begins.

Foliation: a flaky or banded texture pattern in metamorphic rocks.

Fossil: the preserved remains or evidence of life.

Fracture: not breaking into geometric shapes.

Front: the leading edge of an incoming air mass.

Frost action: the breaking of rock material caused by the extreme pressure of freezing water.

Gas giant: a large planet with a low density made mainly of gas.

Geocentric theory: theory that the Earth is the center of the Universe.

Geographic North Pole: a place on the Earth where the axis of rotation meets the ground.

Gibbous: a phase of the Moon in between half and full. Gibbous means egg-shaped.

Glassy (luster): shining like a broken piece of glass; also known as vitreous.

Glassy (texture): a smooth texture in igneous rocks caused by very rapid cooling.

Gnomonic projection: a map projection made by placing a flat piece of paper against a point on the surface of a globe.

GPS: global positioning system. It is a network of satellites that gives an exact position on the surface of the Earth.

Graded bedding: layers of sorted sediments stacked on top of each other.

Gradient: how quickly a value changes across a map.

Great circle: a circle around the Earth that has the core of the Earth at its center.

Greenhouse effect: the trapping of heat in the atmosphere by gases such as carbon dioxide.

Half-life: the amount of time that it takes for 50% of a radioactive element to decay.

Hardness: a mineral's resistance to scratching.

Heliocentric theory: theory that the Sun is the center of the Universe.

Horizontal sorting: when sediments are sorted from side to side as water gradually slows down.

Hot spot: a volcanically active spot that is not on the edge of a plate.

Hurricane: a large organized storm with heavy rains and strong winds.

Hydrologic cycle: see Water cycle.

Hydrosphere: the water portion of the Earth. It includes the oceans, lakes, ice caps, and ground water.

Igneous: rock formed from the solidification of melted minerals.

Index fossil: the fossil of a creature that existed for a short time, but lived over a wide area.

Inference: using information to come to a conclusion or to make a prediction.

Infiltration: water seeping down through soils.

Interpolate: to use two data points to infer a middle value.

Intrusive: rocks solidified while still inside the Earth.

Inverse relationship: the relationship between two numbers where, as one number increases, the other number decreases. Graphs of direct relationships will slant downwards to the right.

Isoline: a line connecting points of equal value.

Isotope: a variety of an element with different numbers of neutrons.

Jovian planet: a large Jupiter-like planet.

Kinetic energy: the energy that comes from an object's motion.

Land breeze: wind that goes from the land to the sea during the night.

Latitude: the degrees north or south of the equator. Latitude lines run side-to-side, and they are all parallel to each other.

Lava: melted rock material that has reached the surface of the Earth.

Leeward: the side of a mountain away from the wind.

Light-year: the distance light travels in a year.

Lithification: the process of turning into solid rock.

Lithosphere: the solid rock portion of the Earth. It includes the crust and the upper mantle.

Longitude: degrees east or west of the Prime Meridian.

Luminosity: the amount of light that a star emits.

Lunar eclipse: the Moon gets darker when the Earth gets in between the Sun and the moon.

Luster: the way a mineral shines or does not shine.

Magma: melted rock material that is still below the surface of the Earth.

Magnetic declination: a measure of how wrong a magnetic compass will be in a certain area of the world.

Magnetic North Pole: a place in extreme northern Canada where all magnetic compasses point to.

Main sequence: a typical star. Stars spend most of their lifetimes as main sequence stars.

Mantle: the deeper layer inside the Earth made of partly melted rock material.

Map projection: a drawing representing a section of the Earth.

Marine air mass: a humid air mass formed over the water.

Mass extinction: the extinction of a larger than normal number of species.

Meander: a bend in a stream.

Meniscus: the downward dip in the surface of water inside a graduated cylinder.

Mercalli Scale: a rating scale that measures the effects of an earthquake.

Meridian: a line of longitude. It runs from the North Pole to the South Pole. All meridians intersect at the poles.

Metallic luster: shining like a polished piece of metal.

Meteor shower: a period when many meteors streak into the atmosphere at once.

Meteor: a streak of light caused by a small particle from space burning up in the atmosphere.

Meteorite: a large meteor that survives the fiery entry through the atmosphere to strike the ground.

Meteoroid: any small particle flying through space.

Meteorology: the study of the atmosphere; primarily the study of weather.

Minute of arc: 1/60 of a degree of angle.

Moho: the boundary between the crust and the asthenosphere.

Moho discontinuity: sudden change in the speed of seismic waves when they cross the Moho.

Mohs scale of hardness: a rating system from one to 10 for measuring a mineral's hardness.

Monomineralic: a rock made up of only one mineral.

Mountain barrier effect: mountains acting as a barrier for rain.

Neap tide: a tide where high tide is not very high and low tide is not very low.

Non-renewable: resources that are gone forever once they are used.

Nor'easter: a low pressure storm where the most intense portion has winds coming out of the northeast.

Nucleus (comet): the icy core of a comet.

Observation: information gathered using any of the five senses. Taking a measurement is also an observation that uses technology to extend the senses.

Occluded front: a combination of a cold front and a warm front.

Oort cloud: a region at the far end of the solar system where most of the comets comc from.

Ore: raw rock material from which resources are extracted.

Orographic effect: air being forced to rise and cool as it crosses a mountain range. Moisture will be forced out of the air as it rises.

Outcrop: an exposed section of bedrock.

Oxbow lake: a meander that was cut off of a stream.

Oxidation: a chemical process where oxygen reacts with a mineral.

Pangaea: ancient super-continent made from most of the Earth's land masses.

Parallax: the apparent shifting of an object's position caused by our movement from one location to another.

Parallelism: the axis of rotation continues to point to the same spot in space as the Earth moves around the Sun.

Parsec: the distance an object will be if it appears to shift by 1 second of arc.

Penumbra: the lighter shadow in an eclipse.

Percent deviation: the calculation of how incorrect a measurement is.

Period of revolution: the amount of time it takes for a planet to travel around the Sun.

Permeability: the ability of water to travel through a material.

Physical weathering: the breaking down of rock material by physical force.

Plane of Earth's orbit: the flat surface that the Earth travels on as it orbits the Sun.

Planet: a large body that orbits the Sun.

Plate boundary: the edge of a solid slab of crust.

convergent plate boundary: place where two plates are crashing into each other.

Plate tectonics: the study of the motions of the plates and within the asthenosphere.

Plutonic: an igneous feature formed inside the Earth.

Polar air mass: a cold air mass.

Polymienrallic: a rock made up of more than one mineral.

Pores: the empty spaces between the sediments in the soil.

Porosity: a measure of how much empty space there is between sediments in a soil.

Porphyry: coarse-textured crystals surrounded by a fine-grained matrix.

Potential energy: the energy that comes from an object's height.

Pressure unloading: the expansion of a rock caused by the release of pressure when a rock gets exposed at the surface of the Earth.

Prevailing wind: the typical direction that the wind will come from in an area.

Prime Meridian: the starting point for longitude. It is a line that goes from the North Pole through Greenwich, England, to the South Pole.

Psychrometer: a device with two

thermometers used for measuring dew point and humidity.

Pulsar: the rapidly spinning core of an exploded star.

P-wave: the primary wave from an earthquake that travels fastest and reaches distant seismic stations first.

Radiation: heat transfer as a form of light.

Radioisotope: a variety of an element that is unstable and will decay.

Rate: how quickly a value changes in time.

Recharge basin: a large pit dug in the ground to give rainwater easy access back into the aquifer.

Red giant: a large, cool star.

Regional metamorphism: a metamorphic change in rock caused by extreme heat and pressure.

Relative humidity: the amount of moisture in the air compared to the amount of moisture the air can hold.

Relative time: placing events in order of which came first and which came second.

Relief: the amount of elevation change in an area. It is "how bumpy an area is."

Residual soil: soil that stays above the bedrock that formed it.

Resinous (waxy): shining like a piece of plastic.

Retrograde motion: moving backwards.

Revolution: an orbit around a star or a planet.

Rhombohedral cleavage: breaking into slanted cubes where the corners do not meet at 90 degrees.

Richter Scale: a rating scale that measures how much energy was released during an earthquake.

Ring of Fire: the nickname given to the Pacific Rim because of all the seismic activity.

Root action: a type of physical weathering caused by roots breaking apart rocks.

Rotational period: the amount of time it takes for a planet to spin.

Saturated: when air is holding as much water as it can.

Scale: the ratio between an object on a map and an object in the real world.

Sea breeze: wind that comes from the sea during the day.

Second of arc: 1/60 of a minute of arc.

Seismic wave: the vibration from an earthquake that travels through the Earth.

Shadow zone: an area on the opposite side of the Earth from an earthquake where no seismic waves are detected. It is a zone that is 102 degrees to 140 degrees away from the earthquake.

Shield volcano: a very wide, gently sloping volcanic mountain made from lava flows.

Sidereal day: a 360-degree turn around an axis. Using a distant star to mark the passage of one day. 23 hours, 56 minutes.

Solar day: marking the passage of one day based on the position of the Sun. 24 hours.

Solar eclipse: the Sun appears to get dark as the Moon gets in between the Earth and Sun.

Source region: the area that an air mass comes from.

Specific heat: a measure of how much energy it takes to heat up one gram of a material.

Spectrum: the different colors of light including those we can't see.

Spring tide: a tide where high tide is very high and low tide is very low.

Spring: a place where groundwater flows out of the ground.

Star trail: the streaking of a star in a photograph caused by Earth's rotation.

Star: a large ball of gases that glows from energy created by fusion.

Stationary front: a place where two air masses meet but neither one is advancing on the other.

Storm surge: a dome of high water level underneath a strong low pressure system.

Streak: the color of powdered mineral.

Sublimation: a phase change from solid directly into gas, skipping the liquid phase. It can also be from gas to liquid.

Supernova: a violent explosion at the end of a star's life.

Superposition: a principle that is used in relative time that places the oldest layer of rock at the bottom.

Suspension: small particles that are carried within flowing water.

S-wave: the secondary wave from an earthquake that travels slower and reaches distant seismic stations second.

Synodic month: a month marked from one Full Moon to the next Full Moon.

Terminator: the line separating day and night.

Terrestrial planet: a small rocky Earth-like planet.

The Principle of Original Horizontality: sedimentary rocks are normally made flat and horizontal.

Till: unsorted sediments of greatly varying sizes that are deposited by glaciers.

Time zone: a section of the Earth approximately 15º in longitude wide where all the clocks are set to the same time.

Time-temperature lag: the time difference between the strongest sunlight and the hottest time.

Topographic map: a map showing the shape of the land using contour lines.

Topography: the shape of the land.

Tornado: a strong, compact storm with high winds.

Trace fossil: the preserved indirect evidence of life, such as footprints and burrows.

Transform plate boundary: place where two plates are sliding past each other.

Transpiration: the process of water vapor leaving the leaves of plants.

Transported soil: soil that was formed in one location and transported to a different location.

Tropical air mass: a warm air mass.

Tropical depression: an organized low pressure system with heavy rain and winds less than 38 mph.

Tropical storm: an organized low pressure system with heavy rain and winds between 39 mph and 73 mph.

Troposphere: the lowest layer of the atmosphere in which all weather takes place.

Tsunami: a large ocean wave caused by an earthquake.

Umbra: the dark portion of the shadow during an eclipse.

Unconformity: a buried erosional surface. It represents a gap in the rock history.

U-shaped valley: a wide valley carved out by glaciers.

Velocity: a fancy word that means speed. Technically, it is a combination of speed and direction.

Vertical sorting: the separating of sediments in size order, with the biggest and heaviest at the bottom and the smallest and lightest at the top.

Vesicle: a bubble in an igneous rock.

Vitreous (glassy): shining like a broken piece of glass.

Volcanic time marker: a layer of volcanic dust that gets deposited almost instantly over a very wide area.

Volcanic: an igneous feature formed on the surface of the Earth.

V-shaped valley: a valley carved by flowing water.

Waning: phases of the Moon where it is getting dimmer from night to night.

Warm front: the leading edge of an incoming warm air mass.

Water cycle: a description of how water recycles itself around the planet.

Water displacement method: finding the volume of an object by measuring the volume of the water it pushes out of a container.

Water dissolution: a type of chemical weathering caused by watering dissolving soluble minerals.

Waxing: phases of the Moon where it is getting brighter from night to night.

Weather: the current condition of the atmosphere.

White dwarf: a small but hot star.

Windward: the side of a mountain that gets hit by wind.

Zenith: the spot in the sky directly above you.

Zone of aeration: the area above the capillary fringe where the pores are filled with air.

Index

D

H

I

K

L

M

N

O

P

Q

R

S

T

About the Author

Phil Medina has degrees in Earth and Space Sciences and Liberal Arts with a concentration in Science, both from Stony Brook University in New York. A resident of South Setauket, Long Island, Phil has enjoyed teaching students in his home school district for 10 years. When not in the classroom, he enjoys hiking and collecting rocks and fossils. Phil is also involved in dog rescue by spending Saturdays at his local shelter as well as fostering homeless dogs in his home until they find good families.